AF347517

# Design and Application of Biomedical Circuits and Systems

# Design and Application of Biomedical Circuits and Systems

Editors

**Alberto Yufera**
**Gloria Huertas**
**Belen Calvo**

MDPI • Basel • Beijing • Wuhan • Barcelona • Belgrade • Manchester • Tokyo • Cluj • Tianjin

*Editors*
Alberto Yufera
Universidad de Sevilla
Spain

Gloria Huertas
Universidad de Sevilla
Spain

Belen Calvo
University of Zaragoza
Spain

*Editorial Office*
MDPI
St. Alban-Anlage 66
4052 Basel, Switzerland

This is a reprint of articles from the Special Issue published online in the open access journal *Electronics* (ISSN 2079-9292) (available at: https://www.mdpi.com/journal/electronics/special_issues/biomedical_circuits_systems).

For citation purposes, cite each article independently as indicated on the article page online and as indicated below:

LastName, A.A.; LastName, B.B.; LastName, C.C. Article Title. *Journal Name* **Year**, *Volume Number*, Page Range.

**ISBN 978-3-0365-0238-0 (Hbk)**
**ISBN 978-3-0365-0239-7 (PDF)**

# Contents


About the Editors . . . . . . . . . . . . . . . . . . . . . . . . . . . . . . . . . . . . . . . **vii**

Preface to "Design and Application of Biomedical Circuits and Systems" . . . . . . . . . . . . . **ix**

**Alberto Yúfera, Gloria Huertas and Belen Calvo**
Design and Application of Biomedical Circuits and Systems
Reprinted from: *Electronics* **2020**, *9*, 1920, doi:10.3390/electronics9111920 . . . . . . . . . . . . . . **1**

**Orazio Aiello**
On the DC Offset Current Generated during Biphasic Stimulation: Experimental Study
Reprinted from: *Electronics* **2020**, *9*, 1198, doi:10.3390/electronics9081198 . . . . . . . . . . . . . **5**

**Tyler Ward, Neil Grabham, Chris Freeman, Yang Wei, Ann-Marie Hughes, Conor Power, John Tudor and Kai Yang**
Multichannel Biphasic Muscle Stimulation System for Post Stroke Rehabilitation
Reprinted from: *Electronics* **2020**, *9*, 1156, doi:10.3390/electronics9071156 . . . . . . . . . . . . . **19**

**Alberto López, Francisco Ferrero, José Ramón Villar and Octavian Postolache**
High-Performance Analog Front-End (AFE) for EOG Systems
Reprinted from: *Electronics* **2020**, *9*, 970, doi:10.3390/electronics9060 . . . . . . . . . . . . . . . . **33**

**Fnu Tala and Benjamin C. Johnson**
MEDUSA: A Low-Cost, 16-Channel Neuromodulation Platform with Arbitrary Waveform Generation
Reprinted from: *Electronics* **2020**, *9*, 812, doi:10.3390/electronics9050812 . . . . . . . . . . . . . **49**

**Stefano Ricci and Valentino Meacci**
FPGA-Based Doppler Frequency Estimator for Real-Time Velocimetry
Reprinted from: *Electronics* **2020**, *9*, 456, doi:10.3390/electronics9030456 . . . . . . . . . . . . . **65**

**Yue Tang, Ronghui Chang, Limin Zhang and Feng Yan**
An Interference Suppression Method for Non-Contact Bioelectric Acquisition
Reprinted from: *Electronics* **2020**, *9*, 293, doi:10.3390/electronics9020293 . . . . . . . . . . . . . **83**

**Hoyoung Yu and Youngmin Kim**
New RSA Encryption Mechanism Using One-Time Encryption Keys and Unpredictable Bio-Signal for Wireless Communication Devices
Reprinted from: *Electronics* **2020**, *9*, 246, doi:10.3390/electronics9020246 . . . . . . . . . . . . . **101**

**Akira Tiele, Alfian Wicaksono, Sai Kiran Ayyala and James A. Covington**
Development of a Compact, IoT-Enabled Electronic Nose for Breath Analysis
Reprinted from: *Electronics* **2020**, *9*, 84, doi:10.3390/electronics9010084 . . . . . . . . . . . . . **111**

**Jaejin Lee, Yangmo Yoo, Changhan Yoon and Tai-kyong Song**
A Computationally Efficient Mean Sound Speed Estimation Method Based on an Evaluation of Focusing Quality for Medical Ultrasound Imaging
Reprinted from: *Electronics* **2019**, *8*, 1368, doi:10.3390/electronics8111368 . . . . . . . . . . . . . **127**

**Bardia Yousefi, Hossein Memarzadeh Sharifipour, Mana Eskandari,**
**Clemente Ibarra-Castanedo, Denis Laurendeau, Raymond Watts, Matthieu Klein and**
**Xavier P. V. Maldague**
Incremental Low Rank Noise Reduction for Robust Infrared Tracking of Body Temperature
during Medical Imaging
Reprinted from: *Electronics* **2019**, *8*, 1301, doi:10.3390/electronics8111301 . . . . . . . . . . . . . . . **137**

**Rifky Ismail, Mochammad Ariyanto, Inri A. Perkasa, Rizal Adirianto, Farika T. Putri,**
**Adam Glowacz and Wahyu Caesarendra**
Soft Elbow Exoskeleton for Upper Limb Assistance Incorporating Dual Motor-Tendon Actuator
Reprinted from: *Electronics* **2019**, *8*, 1184, doi:10.3390/electronics8101184 . . . . . . . . . . . . . . . **151**

**Emanuele Cardillo and Alina Caddemi**
Insight on Electronic Travel Aids for Visually Impaired People: A Review on the
Electromagnetic Technology
Reprinted from: *Electronics* **2019**, *8*, 1281, doi:10.3390/electronics8111281 . . . . . . . . . . . . . . . **171**

# About the Editors

**Alberto Yufera**, Dr., Full Professor was born in Seville (Spain) in 1965. He graduated from Universidad de Sevilla, Spain, in Physics, Electronic field in 1988. He then joined the Departamento de Electrónica y Electromagnetismo, Universidad de Sevilla, later this year as Assistant Professor and, here, obtained his Ph.D. degree in 1994. Since 1991, he has belonged to the Departamento de Tecnología Electrónica, Universidad de Sevilla, where he has served as Full Professor since 2012. In 1989, he became researcher at the Department of Analog Design of the National Microelectronics Canter (CNM), now Institute of Microelectronics at Seville (IMSE). He has participated in research projects financed by the Spanish Government and the European Community. He has published more than 100 technical papers in the main international journals and conferences in the field. His current research interests include analysis and design of analog integrated circuits and systems for signal processing; development of CAD tools for analog circuits, biomedical circuits, and system applications; and bioimpedance-based instrumentation in biological and clinical environments.

**Gloria Huertas** received her Licenciado en Física and Doctor in Ciencias Físicas in 1997 and 2004, respectively, both from Universidad de Sevilla, Spain. Since 1997, she has been with the Departamento de Electrónica y Electromagnetismo, Universidad de Sevilla, Spain, where she is Full Professor, as well as with the Instituto de Microelectrónica de Sevilla, Centro Nacional de Microelectrónica, Sevilla, Spain. Her current research interests are in the development of alternative bio-instrumentation circuits and systems required for classical reproduction and to propose new measurement techniques at bio-medical labs to improve the quality of acquired biosignals. She has published a book, several book chapters, and more than sixty articles in national and international journals and conferences in her scientific field. She has participated in research and development projects at both national and European level and in several industrial contracts as well as in international development cooperation projects. She also serves as a scientific reviewer for several national and international journals and conferences. It is also worth noting that she has four invention patents.

**Belen Calvo** received his B.Sc. degree in Physics and Ph.D. degree in Electronic Engineering from the University of Zaragoza, Zaragoza, Spain, in 1999 and 2004, respectively. She is currently Senior Researcher with the Group of Electronic Design, Aragon Institute for Engineering Research (GDE-I3A), University of Zaragoza. Her current research interests include analog and mixed-mode CMOS IC design, on-chip programmable circuits, low-voltage low-power monolithic sensor interfaces, and wideband front-end IC transceivers.

# Preface to "Design and Application of Biomedical Circuits and Systems"

Since the appearance of electronics as an enabling technology, circuits and systems have been developed to improve biomedical measurements and design new medical equipment for specific tests in the health field. Diagnosis, therapy, clinical testing, and bio-signal monitoring between other medical tasks are properly and accurately performed using electronic equipment and, today, are considered vital for medical data acquisition tasks. The development of new sensing technologies, biomaterials, microelectronic devices, microfluidic systems, micro-electro-mechanical systems (MEMs), etc., open the window to new biomedical circuits and systems opportunities to measure "better" and also using "alternative" methods to find the relevant information required by teams of physicians and biologists. This Special Issue is mainly devoted to incorporating proposals of bio-sensing signals based on new circuits and systems approaches. In general, it is focused on new bio-signal analog front-end (AFE) circuits; specific circuits development for known and new sensor/sensing approaches; circuits for biomedical signal processing; low-voltage and low-power circuits and their application to implants and wearable devices; circuits and systems for clinical applications; circuits for sensing/actuation in MEM systems, Lab-on-a-Chip (LoC), micro-total-analysis Systems (uTAS); and cell assays and manipulation. Other main topics covered in this Special Issue include circuits for bioimpedance test; capacitive-based circuits, circuits for new sensing devices and microeletrodes; ECG, EEG, EMG, EoG, etc. circuits and systems; circuits for implants and wearable devices; LP/LV circuits in biomedical environments; circuits and systems in clinical applications; circuits for cell, DNA, bacteria, virus assays; and brain interfaces.

**Alberto Yufera, Gloria Huertas, Belen Calvo**
*Editors*

 *electronics*

*Editorial*

# Design and Application of Biomedical Circuits and Systems

**Alberto Yúfera [1,*], Gloria Huertas [1] and Belen Calvo [2]**

[1] Instituto de Microelectrónica de Sevilla (IMSE-CSIC), Universidad de Sevilla, Av. Americo Vespuccio 24, 41092 Sevilla, Spain; gloria@imse-cnm.csic.es

[2] Grupo de Diseño Electrónico, Instituto de Investigación en Ingeniería de Aragón (GDE-I3A), Universidad de Zaragoza, 50009 Zaragoza, Spain; becalvo@unizar.es

* Correspondence: yufera@imse-cnm.csic.es

Received: 5 November 2020; Accepted: 9 November 2020; Published: 15 November 2020

## 1. Introduction

The development of new sensing technologies, biomaterials, microelectronic devices, microfluidic systems and micro-electro-mechanical systems (MEMs) etc., opens the window to new biomedical circuits and system opportunities to measure "better", and to develop "alternative" methods to find relevant information for physician and biologist teams, in applications such as diagnosis, therapy, clinical tests and bio-signal monitoring. However, the accomplishment of new medical equipment for specific tests in the health field poses significant challenges regarding the electronic circuits and systems needed, whose performance is vital for proper and accurate data acquisition tasks.

## 2. Design and Application of Biomedical Circuits and Systems

This Special Issue is devoted mainly to incorporating proposals of bio-sensing signals based on new circuits and system approaches. In general, it is focused on new bio-signal analog front-end (AFE) circuits; the development of specific circuits for known and new sensor/sensing approaches; circuits for biomedical signal processing; low-voltage and low-power (LV/LP) circuits and its application to implantable and wearable devices; circuits and systems for clinical applications; circuits for sensing/actuation in MEM systems, lab-on-a-chip (LoC), micro-total-analysis systems (uTAS); cell assays and manipulation, etc. Main topics of interest are well described by the Special Issue keywords (but not limited to):

- Analog front-end (AFE) circuits;
- Circuits for bioimpedance test;
- Capacitive based circuits;
- Circuits for new sensing devices and microelectrodes;
- ECG, EEG, EMG, EoG etc. circuits and systems;
- Circuits for implantable and wearable devices;
- LP/LV circuits in biomedical environments;
- Micro-energy harvesting;
- Circuits and systems in clinical applications;
- Circuits for cells, DNA, bacteria, viruses etc. assays;
- Brain interfaces;
- Internet of Things for remote healthcare;

In the present Special Issue, twelve papers have been successfully incorporated. We hope you enjoy reading this Special Issue and are inspired to address the technological challenges to help the medical industry and biologists to increase the human quality of life, which is the main objective. These are the contribution papers:

1. On the DC Offset Current Generated during Biphasic Stimulation: Experimental Study [1].
2. Multichannel Biphasic Muscle Stimulation System for Post Stroke Rehabilitation [2].
3. High-Performance Analog Front-End (AFE) for EOG Systems [3].
4. MEDUSA: A Low-Cost, 16-Channel Neuromodulation Platform with Arbitrary Waveform Generation [4].
5. FPGA-Based Doppler Frequency Estimator for Real-Time Velocimetry [5].
6. An Interference Suppression Method for Non-Contact Bioelectric Acquisition [6].
7. New RSA Encryption Mechanism Using One-Time Encryption Keys and Unpredictable Bio-Signal for Wireless Communication Devices [7].
8. Development of a Compact, IoT-Enabled Electronic Nose for Breath Analysis [8].
9. A Computationally Efficient Mean Sound Speed Estimation Method Based on an Evaluation of Focusing Quality for Medical Ultrasound Imaging [9].
10. Incremental Low Rank Noise Reduction for Robust Infrared Tracking of Body Temperature during Medical Imaging [10].
11. Soft Elbow Exoskeleton for Upper Limb Assistance Incorporating Dual Motor-Tendon Actuator [11].
12. Insight on Electronic Travel Aids for Visually Impaired People: A Review on the Electromagnetic Technology [12].

We are conscious about the very wide scope of biomedical circuits and systems applications, and that our contribution it is only a grain of sand more, but we expect to be it useful for knowledge progress in the field.

## 3. Conclusions

Biomedical engineering is today one of the most important research fields in the world. This fact is parallel with the health challenges surrounding the improvement of human health as one of the main vehicles to increase the quality of life. The maturity of many technologies, such as microelectronic, biomaterial, microfluidic, together with progress in the biology and medicine fields, develop alternative solutions for medical evaluation, diagnosis, therapy and research in general, opening the opportunity for new medical devices, e.g., lab-on-a-chip, wearable technology, and implants. The biomedical electronic industry supports the development of many of these new devices, as the main technologies for bio-signal acquisition, processing, and communication. In this Special Issue contribution, we present some significant contributions for biomedical applications, which, of course, should be fulfilled and improved by future contributions.

**Author Contributions:** Conceptualization, A.Y., G.H. and B.C.; methodology, A.Y., G.H. and B.C.; writing—review and editing and project administration, A.Y., G.H. and B.C. All authors have read and agreed to the published version of the manuscript.

**Funding:** This research received no external funding.



## References

1. Aiello, O. On the DC Offset Current Generated during Biphasic Stimulation: Experimental Study. *Electronics* **2020**, *9*, 1198. [CrossRef]
2. Ward, T.; Grabham, N.; Freeman, C.; Wei, Y.; Hughes, A.-M.; Power, C.; Tudor, J.; Yang, K. Multichannel Biphasic Muscle Stimulation System for Post Stroke Rehabilitation. *Electronics* **2020**, *9*, 1156. [CrossRef]
3. López, A.; Ferrero, F.; Villar, J.R.; Postolache, O. High-Performance Analog Front-End (AFE) for EOG Systems. *Electronics* **2020**, *9*, 970. [CrossRef]

4.  Tala, F.; Johnson, B.C. MEDUSA: A Low-Cost, 16-Channel Neuromodulation Platform with Arbitrary Waveform Generation. *Electronics* **2020**, *9*, 812. [CrossRef]

5.  Ricci, S.; Meacci, V. FPGA-Based Doppler Frequency Estimator for Real-Time Velocimetry. *Electronics* **2020**, *9*, 456. [CrossRef]

6.  Tang, Y.; Chang, R.; Zhang, L.; Yan, F. An Interference Suppression Method for Non-Contact Bioelectric Acquisition. *Electronics* **2020**, *9*, 293. [CrossRef]

7.  Yu, H.; Kim, Y. New RSA Encryption Mechanism Using One-Time Encryption Keys and Unpredictable Bio-Signal for Wireless Communication Devices. *Electronics* **2020**, *9*, 246. [CrossRef]

8.  Tiele, A.; Wicaksono, A.; Ayyala, S.K.; Covington, J.A. Development of a Compact, IoT-Enabled Electronic Nose for Breath Analysis. *Electronics* **2020**, *9*, 84. [CrossRef]

9.  Lee, J.; Yoo, Y.; Yoon, C.; Song, T.-k. A Computationally Efficient Mean Sound Speed Estimation Method Based on an Evaluation of Focusing Quality for Medical Ultrasound Imaging. *Electronics* **2019**, *8*, 1368. [CrossRef]

10. Yousefi, B.; Sharifipour, H.M.; Eskandari, M.; Ibarra-Castanedo, C.; Laurendeau, D.; Watts, R.; Klein, M.; Maldague, X.P.V. Incremental Low Rank Noise Reduction for Robust Infrared Tracking of Body Temperature during Medical Imaging. *Electronics* **2019**, *8*, 1301. [CrossRef]

11. Ismail, R.; Ariyanto, M.; Perkasa, I.A.; Adirianto, R.; Putri, F.T.; Glowacz, A.; Caesarendra, W. Soft Elbow Exoskeleton for Upper Limb Assistance Incorporating Dual Motor-Tendon Actuator. *Electronics* **2019**, *8*, 1184. [CrossRef]

12. Cardillo, E.; Caddemi, A. Insight on Electronic Travel Aids for Visually Impaired People: A Review on the Electromagnetic Technology. *Electronics* **2019**, *8*, 1281. [CrossRef]

**Publisher's Note:** MDPI stays neutral with regard to jurisdictional claims in published maps and institutional affiliations.

*Article*

# On the DC Offset Current Generated during Biphasic Stimulation: Experimental Study

**Orazio Aiello**

Department of Electrical and Computer Engineering, National University of Singapore, Singapore 117583, Singapore; orazio.aiello@nus.edu.sg

Received: 30 June 2020; Accepted: 22 July 2020; Published: 25 July 2020

**Abstract:** This paper deals with the DC offset currents generated by a platinum electrode matrix during biphasic stimulation. A fully automated test bench evaluates the nanoampere range DC offset currents in a realistic and comprehensive scenario by using platinum electrodes in a saline solution as a load for the stimulator. Measurements are performed on different stimulation patterns for single or dual hexagonal stimulation sites operating simultaneously and alternately. The effectiveness of the return electrode presence in reducing the DC offset current is considered. Experimental results show how for a defined nominal injected charge, the generated DC offset currents differ depending on the stimulation patterns, frequency, current amplitude, and pulse width of a biphasic signal.

**Keywords:** DC offset current; electrode stimulation; biphasic signal; platinum electrode matrix

---

## 1. Introduction

Neural stimulation shows reliable effectiveness in concrete cures where traditional medication is not as effective at reducing tremors in epilepsy or in restoring the sight in a person affected by retinitis pigmentosa. To elicit nerve activities, tissues are usually excited by zero-net charge biphasic current pulses [1], whose features are reported in Figure 1. The first cathodic pulse usually elicits the desired neural response, while the second pulse only aims to neutralize charge across the stimulated tissues. A perfect charge balance in the neural stimulator, such as cochlear or retinal integrated circuit (IC) implants, is required to avoid DC currents or electrode resting potentials that can generate toxic species and, in turn, damage tissues [2–11].

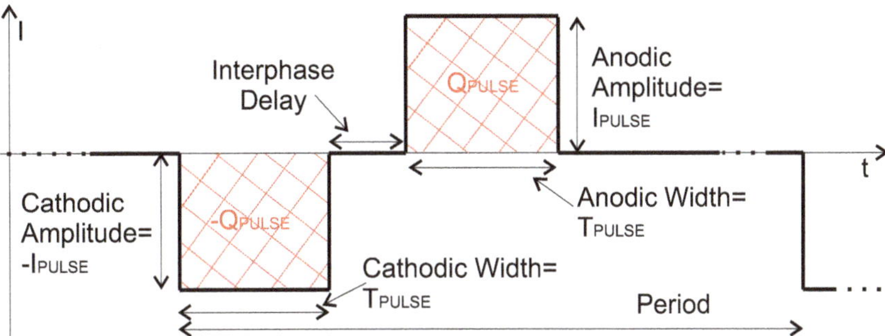


**Figure 1.** Pictorial representation of a biphasic current stimulation. Typical electric characteristics are highlighted.

The charge during the opposite phases of sourcing and sinking currents should be balanced, as well as the impedance of the electrodes needs to be perfectly matched. For this purpose, several techniques and stimulation strategies for a precise charge balancing have been proposed at the IC level to minimize the undesired DC offset current generated by the biphasic stimulation [12–18].

During a biphasic stimulation, each platinum electrode in the saline bath defines an electrode-electrolyte interface while a charge is injected through faradaic and non-faradaic charge-transfer mechanisms. These can lead to changes in the surface chemistry of the platinum electrodes under study and are the source of DC offset currents originating at the electrode-electrolyte interface [19]. On this basis, IC designers for biomedical devices have shown a relevant effort to reduce any mismatch between the current sources that provide the biphasic pulse at the electrodes-tissue interface. The more macroscopic each electrode-tissue interface's size, the more significant is the mismatch over time and between implantees. This represents a bottleneck in the DC offset current generation. Based on the weakness of the signal to monitor (tens of nanoamperes), the DC offset current value in the literature usually refers to CAD simulation or measurements where the electrodes and the tissues are substituted with an equivalent quasi-static loss electric model constituted by passive components (resistor in series with a parallel capacitor and resistor). Moreover, such a DC offset current value reported in the literature relates to a defined stimulation pattern and a specific set of electrical parameters of the biphasic signal [12–18].

In this paper, the DC offset currents generated by a neural IC implant that provides the current stimulation through a platinum electrode matrix placed in a saline bath are investigated. This is because measurement results performed with physiological saline emulate at best the real case where the electrodes interface with tissues in neural implantation. The most common hexagonal stimulation patterns operating in different conditions of biphasic pulses' amplitude and duration have been analyzed. An automatic measurement setup aimed to monitor DC offset current is described in Section 2. Then, the procedure for a reliable data acquisition is shown in Section 3. In Section 4, the measurement results according to a different frequency, amplitude, and duration of the biphasic pulses and for different stimulation patterns are shown. The DC offset currents of two hexagonal stimulation sites operating simultaneously or stand-alone are investigated. In Section 5, the conclusions are drawn.

## 2. Measurement Setup

### 2.1. System Overview

The purpose of the measurement test bench is to monitor the DC offset currents generated by platinum electrodes in a saline bath when they are actively driven by a neural stimulator. This is to validate the safety of the overall system (from the neural IC implant to the electrode matrix), as well as the chosen stimulation regime. The sketch of the DC offset current measurement test bench and the respective photo are shown in Figure 2.

Such a test bench is remotely controlled by a personal computer (PC in Figure 2) running an automated test bench (USB controlled by Matlab) aimed to measure DC offset currents through a statistical approach based on data storage redundancy to achieve valuable results. Notice that because of the weak nature of the signal to be monitored, the measurement setup is placed into a metal enclosure that acts as a Faraday cage. Such metal enclosures help in minimizing power line interference, external noise, and any rapid drift in temperature and airflow during the measurements. Figure 3 shows the content of the Faraday cage. A graphic interface defines the stimulation pattern sending the related protocol to the telemetry module (left side in Figure 3). This, in turn, sets the electric specifications of the biphasic current pulses of the IC stimulator through a fully differential two-wire interface. The IC neural stimulator employed in the investigation reported in this paper [18] is placed into a specific module test board (bottom-center in Figure 3). Each of the biphasic currents delivered by the neural stimulator flow through a sense resistor mounted on the test board to the respective electrode in a saline bath (bottom-right in Figure 3).

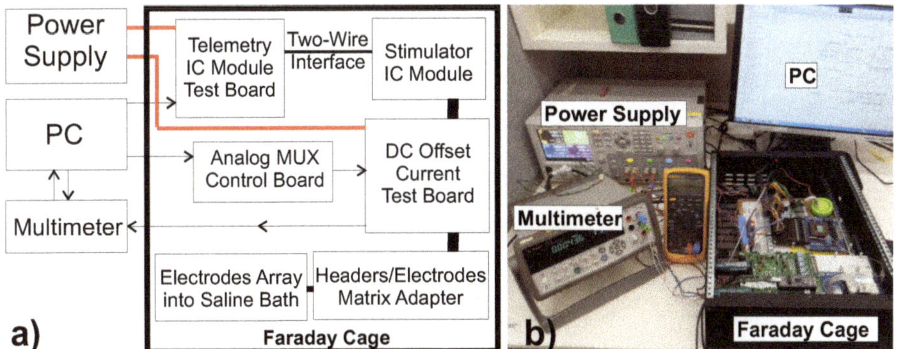


**Figure 2.** (**a**) Sketch of the DC offset current measurement test bench. (**b**) Respective photo.

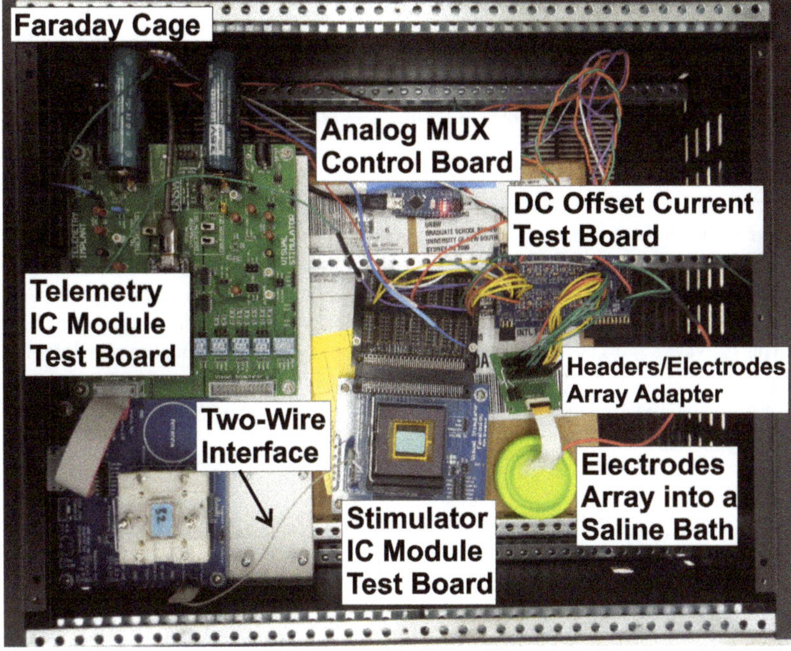


**Figure 3.** DC offset current measurement setup inside the metal enclosure as depicted in Figure 2.

## 2.2. DC Offset Current Detection Test Board

The test board represented in Figure 4 allows the detection of the DC offset currents by means of the voltages produced by each biphasic current pulse flowing through a sense resistor. Indeed, such voltages are low pass filtered before being amplified by a cascade of two low offsets, low input bias current instrumentation amplifiers (Linear Technology, LT1167 [20]) for each channel (Figure 4). Since the system needs to measure DC offset current lower than 100 nA across a sense resistor of 100 $\Omega$, corresponding to a voltage drop of 10 $\mu$V, two gain stages are used. In order to circumvent the increase in required PCB area and system cost, the final design consists of 16 copies of the instrumentation amplifier. The outputs from such amplifiers are subsequently multiplexed to a second stage instrumentation amplifier with a gain of 100 via an analog multiplexer (ADG426 [21]). The overall system can evaluate current adequately down to 10 nA. The acquired data are sent to the microcontroller (analog MUX control board in Figures 2a and 3) via USB. Such a microcontroller drives the analog multiplexer on the PCB where the above-mentioned matrix of sense resistance and the cascade gain stage are placed. The output of the analog multiplexer is measured by a multimeter (Agilent 34411A) that is interfaced with the personal computer via USB (as in Figure 2).

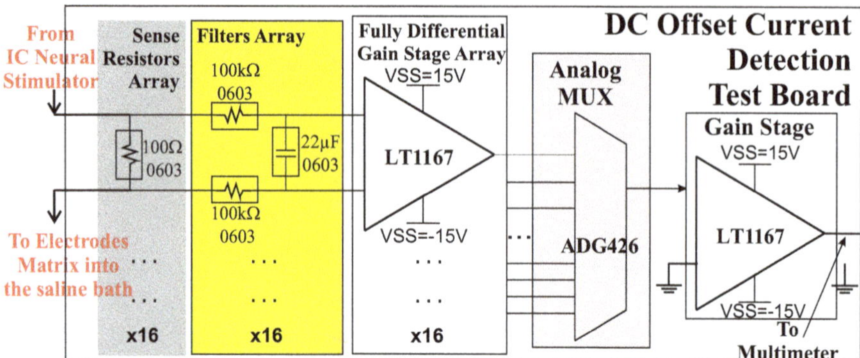


**Figure 4.** Sketch of the DC offset current detection test board.

### 2.3. Platinum Electrodes Matrix

A matrix consisting of 24 electrodes (Figure 5a [22]) is placed into a saline bath with another single bigger electrode as in Figure 5b. Both are inserted and isolated (through silicone) into a slot on the lid of a sample jar containing the artificial physiological saline solution. The bigger electrode is employed as the monopolar return used to assess the quasi-monopolar stimulations pattern described in Section 3. The platinum electrodes are hermetically enclosed in a physiological saline bath using silicone. This is to keep the same ion concentration in the saline bath so that the physiological saline does not need to be replaced every new day of measurements. Moreover, thanks to the silicone's presence, a reciprocal distance between the electrode matrix and the return electrode (≈1 cm) is set.

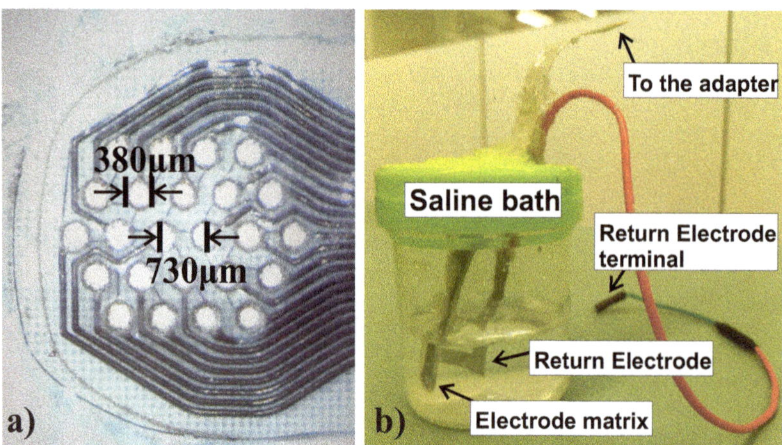


**Figure 5.** (**a**) Photo of the 24 electrode matrix (E24015) used in the DC offset current measurement setup [22]. (**b**) Saline bath.

## 3. Stimulation Procedures And Methods

### 3.1. Preliminary Procedures

Before providing the biphasic stimulation and running the DC offset current measurements, a robust setup has to be built taking into account the critical interfacing with the electrodes in a saline bath. For this reason, the electrode matrix was carefully cleaned and connected to the neural IC implant according to the following procedures. First, a drop of 80% ethanol solution is applied on the electrode matrix and to facilitate the gentle removal of the accumulated particles (if any) on the surface of the electrodes with the tip of a tweezer. Indeed, particles can lay on the surface of the electrodes as a result of ionic impurities within the saline solution (i.e., silicone employed for the electrode matrix and return electrode (RE) enclosure in the saline bath) after long runs of measurements with the same saline bath. After this first cleaning, the electrode matrix is rinsed and left in deionized water for a few hours before starting DC offset current measurements. Then, the electrode matrix is immersed into the saline bath and connected to the respective terminals through an adapter. This adapter is then connected to the headers of the DC offset current detection test board. As a final check of the connections of the entire measurement setup (neuro IC implant, DC offset current test board, header/electrodes adapter, and electrode matrix in the saline bath), the biphasic pulse is visualized with a scope. The probe-scope is then removed during the measurement procedures. To avoid current flowing into the diode of the amplifiers, the voltage supply dynamic range of the amplifier on the DC offset current test board has to be higher (±15 V) than the voltage dynamic range of the neural stimulator (12 V).

### 3.2. Stimulation Patterns

The two most common retinal stimulation patterns employed to restore the sight in visually impaired persons are sketched in Figure 6 where they are named as the All Adjacent (AA) and the Hex Quasi-monopolar (HQ) return configurations. In both configurations, the electrodes are arranged in a hexagonal pattern that represents the elementary stimulation site. In the All Adjacent configuration (Figure 6a), the current flows to the tissues from the center of the hexagon, and the current return path is through the six surrounding electrodes [23]. This stimulation arrangement ensures that the stimulation current flow is localized to each hex center, but at the same time, such localized current flow reduces the current flow to excitable tissue when compared with stimulating against a distant return electrode. For this reason, an improved stimulation approach is represented by the hex quasi-monopolar configuration sketched in Figure 6b. In this approach, in addition to the localized hexagon site, the current can also flow to a distant, large, and low-impedance monopolar return electrode located on the stimulating module capsule. This arrangement allows for a decrease in the power drawn by the system [24].

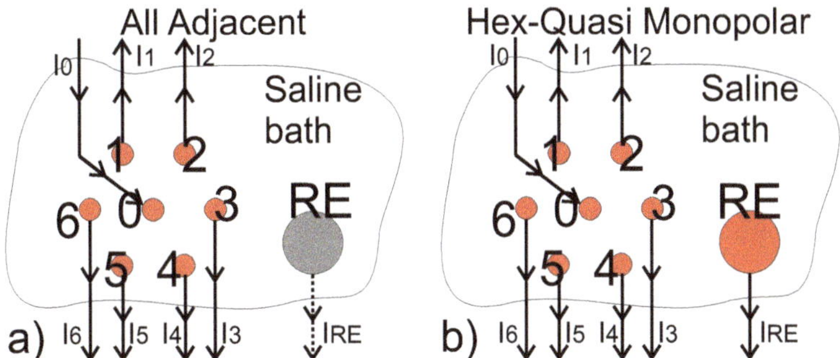


**Figure 6.** Sketch of (**a**) All Adjacent (AA) and (**b**) Hex Quasi-monopolar (HQ) return stimulation patterns with the respective current pulses' directions and the return electrode (RE).

### 3.3. Stimulation Specifications

The maximum current amplitude and pulse duration of the biphasic stimulation is related to the charge injection limit (maximum value of 210 $\mu C/cm^2$) [3]. This results in a maximum charge delivery of 238 nC (per electrode and per phase) for electrodes with an opening of 380 $\mu m$ in diameter. Note that the geometric surface area is conservatively used in the determination of electrode charge-carrying capacity. On this basis, the maximum pulse duration is set equal to $T_{PULSE_{MAX}} = 500$ $\mu s$, and the maximum current capability of the stimulator under test is limited to $I_{PULSE_{MAX}} = 460$ $\mu A$. Therefore, the maximum charge delivery per phase is equal to:

$$Q_{PULSE_{MAX}} = I_{PULSE_{MAX}} \cdot T_{PULSE_{MAX}} = 230 \text{ nC} \tag{1}$$

and just lower than the above-mentioned value. The interphase delay is equal to 5 $\mu s$ as an IC stimulator constraint. The relevant stimulation frequencies of 50 Hz and 25 Hz were considered.

### 3.4. Measurements Procedures

An automated test bench developed to collect the DC offset operates according to the following steps. First, the neural stimulator is turned off to measure the offset of the amplification chain. The offset voltage of each cascade gain stage is stored, and the measurement is repeated 100 times to find the average value of the offset voltage for each channel. Then, the neural IC implant is turned on according to a defined pattern. The acquisition system waits for two minutes before starting to collect data to avoid any starting drift phenomena. This procedure is repeated 100 times, and an average voltage value is calculated for each channel. Subtracting such data from the respective offset, the voltage drop due to DC offset current and to the test bench is calculated.

The investigation is focused on the trend of DC offset current generation with different biphasic signal features like frequency, current amplitude, pulse width, and the reciprocal interaction among stimulation sites. Because of the weak nature of the signal to detect, the measurement results are reported considering average values and the respective range of variability equal to $\pm 3\sigma$ where $\sigma$ is the standard deviation.

## 4. Measurement Results

The DC offset current generated by platinum electrodes during biphasic stimulation in a saline bath is investigated firstly referring to the All Adjacent and the Hex Quasi-monopolar stimulation patterns for a single excitation site. Based on these results, the investigation will be focused on the Hex Quasi-monopolar stimulation pattern considering two hexagonal excitation sites operating simultaneously and with a stimulation frequency of 25 Hz and 50 Hz. A sweep of current amplitude $I_{PULSE}$ for a constant pulse width $T_{PULSE}$ is first considered. Then, a sweep of the pulse width $T_{PULSE}$ to a constant current amplitude $I_{PULSE}$ is reported. In this way, for a charge pulse nominally equal to $Q_{PULSE} = I_{PULSE} \cdot T_{PULSE}$, a different contribution of the current amplitude $I_{PULSE}$ and of the pulse width $T_{PULSE}$ on the overall DC offset current generation can be highlighted. As a final analysis, such a DC offset current generated by two hexagonal excitation sites operating simultaneously is compared with the sum of the DC offset currents of each hexagonal site operating stand-alone with the return electrode.

### 4.1. All Adjacent and Hex Quasi-Monopolar Stimulation Patterns' Comparison

As depicted in Figure 6a,b respectively for the All Adjacent and the Hex Quasi-monopolar stimulation patterns, the stimulation current $I_{PULSE}$ flows into the saline bath from the central electrode (labeled 0) to the surrounding ones (labeled 1 to 6) and to the return electrodes (REs). The math relation among the stimulation current for each electrode can be written as:

$$I_{PULSE} = I_0 = - \left( \sum_{j=1}^{6} I_j + I_{RE} \right) \tag{2}$$

Thus, accordingly, the respective DC offset current of the central electrode $i_0$ is split into the others:

$$i_0 = - \left( \sum_{j=1}^{6} i_j + i_{RE} \right) \tag{3}$$

Therefore, the DC offset current from the central electrode $i_0$ is much higher than the others for each hexagon excitation site. In Figure 7, this current $i_0$ and the sum of the others versus the biphasic current amplitude at the pulse width $T_{PULSE} = T_{PULSE_{MAX}} = 500$ μs are reported. Measurements were performed on a single stimulation site for both the All Adjacent and the Hex Quasi-monopolar stimulation patterns. Since the DC offset currents come from a relative mismatch between the anodic and cathodic charges' path in the stimulator, it increases with the current stimulation amplitude. These measurements highlight how a Hex Quasi-monopolar stimulation pattern offers a reduced DC offset current compared with the All Adjacent configuration. This is due to the role of the return electrode (RE). In such a bigger electrode, current flows from all the active hexagons site belonging to the electrode matrix. This occurs even in the All Adjacent return configuration that nominally should not require the return electrode. In fact, at the end of each biphasic period, all the electrode terminals are shorted together. Thus, because of its bigger size, the return electrode has a higher residual charge flowing in it. As a consequence, the return electrode presence offers a common path to flow for the DC offset currents coming from the electrode matrix. This, in turn, implies a reduction of the DC offset current spreading across the electrodes in the matrix. For this reason, the Hex Quasi-monopolar pattern in which the return electrode is actively involved in the stimulation shows a reduced DC offset current as reported in Figure 7. On this basis, the Hex Quasi-monopolar stimulation pattern is the preferred one in a real experiment and further considered in the following.

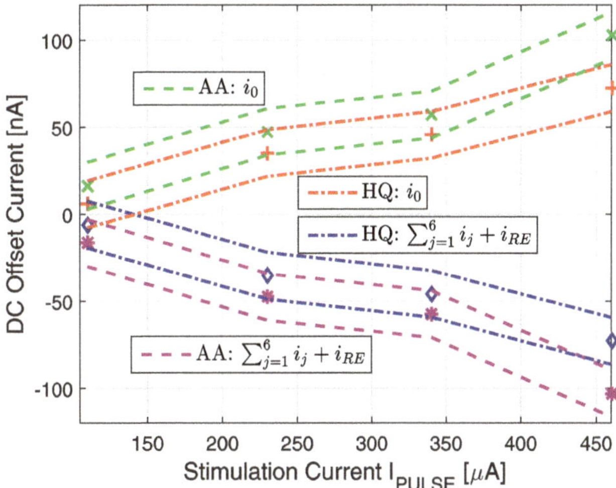


**Figure 7.** Biphasic current amplitude versus DC offset current ($\pm 3\sigma$ range around the average value defined by the dashed and dash-dotted lines) for All Adjacent (AA: $\times$ for the central current $i_0$; $\star$ for the sum of the other surrounding currents) and Hex Quasi-monopolar (HQ: + for the central current $i_0$; $\diamond$ for the sum of the other surrounding currents) stimulation patterns at 50 Hz. Single hexagon (HEXa). Pulse width constant equal to $T_{PULSE} = T_{PULSE_{MAX}} = 500$ μs.

### 4.2. Hex Quasi-Monopolar Multi-Site Excitation

In a practical case, the electrode matrix is made of tens of multiple hexagonal stimulation sites alternately activated that share the same return electrode [12–18]. Figure 8, representing two hexagonal stimulation sites HEXa and HEXb that are simultaneously active and share the same return electrode. The mathematical relationship among the stimulation currents $I$ for a Hex Quasi-monopolar stimulation pattern is:

$$2 \cdot I_{PULSE} = I_{0a} + I_{0b} = - \left( \sum_{j=1}^{6} (I_{ja} + I_{jb}) + I_{RE} \right) \tag{4}$$

so that for the respective DC offset currents, $i$ can be written as:

$$i_{0a} + i_{0b} = - \left( \sum_{j=1}^{6} (i_{ja} + i_{jb}) + i_{RE} \right) \tag{5}$$

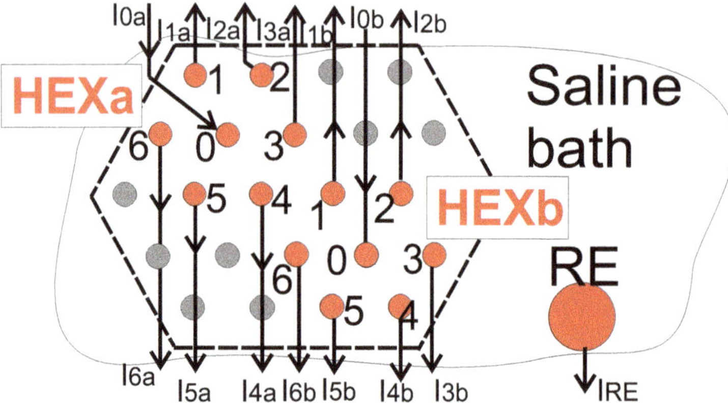


**Figure 8.** Sketch of a Hex Quasi-monopolar (HQ) return stimulation pattern using two active hexagon (HEXa and HEXb) according to the electrode matrix shown in Figure 5.

The sum of the DC offset currents of the central electrodes for two hexagonal sites $i_{0a} + i_{0b}$ into their respective standard deviation range has to match in absolute value with the sum of the DC offset currents of the other electrodes. Nevertheless, not only the DC offset currents of such a central electrode, but even those of every electrode involved in the stimulation were acquired. This was to check the effectiveness of the measurement system and, thus, the validity of the acquired data. After having considered this in every measurement run, for simplicity and clarity, only the DC offset currents of the central electrodes are reported in the following measurement results. The current amplitude $I_{PULSE}$ of each hexagonal site was swept (from $I_{PULSE_{MAX}}/4$ to $I_{PULSE_{MAX}}$) keeping the same pulse duration $T_{PULSE_{MAX}} = 500$ μs as shown in Figure 9a for a biphasic signal frequency of 50 Hz (range among the dashed red lines) and 25 Hz (range among the dash-dotted blue lines). Similarly, the pulse width $T_{PULSE}$ was swept (from $T_{PULSE_{MAX}}/4$ to $T_{PULSE_{MAX}}$) keeping the same pulse duration $I_{PULSE_{MAX}} = 460$ μA as shown in Figure 9b for a biphasic signal frequency of 50 Hz and 25 Hz. This is to highlight the respective contribution to the overall DC offset current spread into the saline. In both of these measurement results, stimulation frequencies of 50 Hz and 25 Hz are reported. The comparison of these results highlights how lowering the frequency of the biphasic signal reduces the DC offset current. This implies that the mismatch between the anodic and cathodic charge pulses depends on their timing to reach the nominal stimulation current values $-I_{PULSE}$ and $I_{PULSE}$ in the two stimulation phases. Notice that in Figure 9a,b, the absolute mismatch increases with an increased absolute charge entity $Q_{PULSE} = I_{PULSE} \cdot T_{PULSE}$ during the stimulation. However, lowering the amplitude of the current pulses $I_{PULSE}$ for a given pulse duration $T_{PULSE}$ (as in Figure 9a) reduces the DC offset current more than lowering the pulse duration $T_{PULSE}$ for a given current pulse amplitude $I_{PULSE}$ (as in Figure 9b). To better highlight which one among current amplitude $I_{PULSE}$ and pulse width $T_{PULSE}$ contributes more to the DC offset current generation, the same data reported in Figure 9a,b are translated in Figure 10 into the respective nominal charge provided during biphasic stimulation, referring to a 50 Hz frequency.

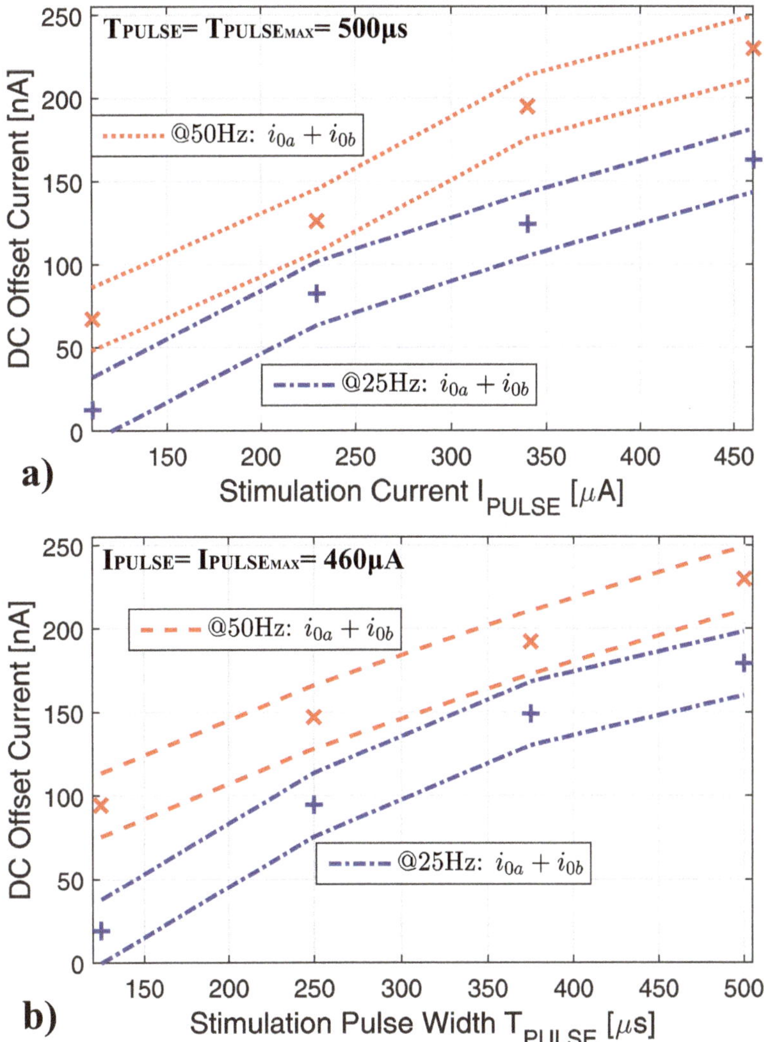


**Figure 9.** Hexagon HEXa and HEXb excitations according to the stimulation pattern in Figure 8. (**a**) Biphasic current amplitude $I_{PULSE}$ versus DC offset current ($\pm 3\sigma$ range around the average value defined by dashed and dash-dotted lines) for the Hex Quasi-monopolar (HQ) stimulation pattern at 50 Hz (see $\times$) and 25 Hz (see +). Pulse width constant equal to $T_{PULSE} = T_{PULSE_{MAX}} = 500$ µs. (**b**) Biphasic pulse duration $T_{PULSE}$ versus DC offset current ($\pm 3\sigma$ range around the average value defined by dashed and dash-dotted lines) for the Hex Quasi-monopolar (HQ) stimulation pattern at 50 Hz (see $\times$) and 25 Hz (see +). Pulse amplitude constant equal to $I_{PULSE} = I_{PULSE_{MAX}} = 460$ µA.

In Figure 10, the nominal injected charge $Q_{PULSE} = I_{PULSE} \cdot T_{PULSE}$ versus the DC offset current for a constant stimulation current $I_{PULSE} = I_{PULSE_{MAX}} = 460$ µA (range among the dashed red lines for the sum of the two central electrodes' DC offset currents $i_{0a} + i_{0b}$) and for a constant pulse width $T_{PULSE} = T_{PULSE_{MAX}} = 500$ µs (range among the dash-dotted black lines for the sum of the two central electrodes' DC offset currents $i_{0a} + i_{0b}$). The comparison shows how for a given injected charge $Q_{PULSE} = I_{PULSE} \cdot T_{PULSE}$, the DC offset current is reduced with a shortened pulse width $T_{PULSE}$ and choosing the current amplitude $I_{PULSE}$ accordingly. As a result, the stimulation timing both in terms of frequency and pulse width amplitude is the most critical in DC offset current generation. This suggests how the offset generation is related to electrophysiology phenomena at the platinum electrodes' interface during the charge pulses.

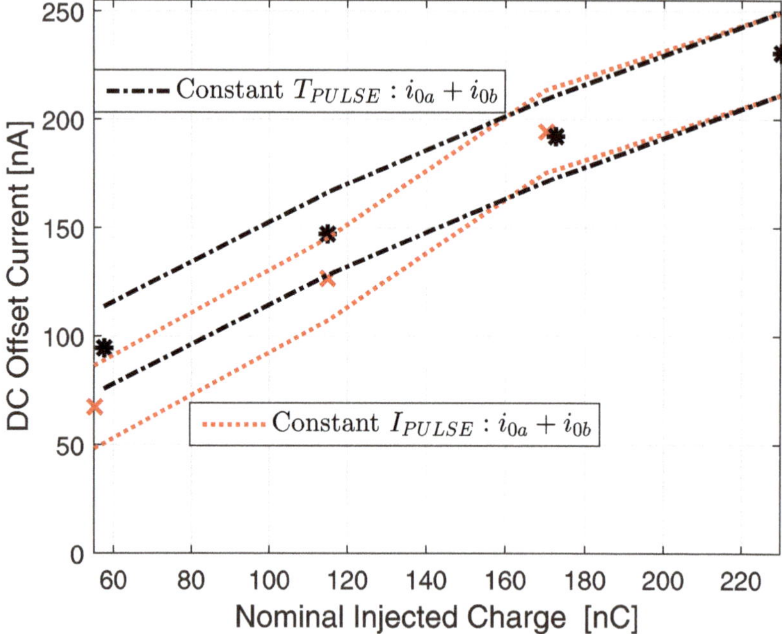


**Figure 10.** Hexagon HEXa and HEXb excitations according to the stimulation pattern in Figure 8. Nominal biphasic charge $Q_{PULSE}$ versus DC offset current ($\pm 3\sigma$ range around the average value defined by dashed and dash-dotted lines) for a constant current pulse $I_{PULSE} = I_{PULSE_{MAX}} = 460$ µA (dashed red lines around ×) and for constant pulse duration $T_{PULSE} = I_{PULSE_{MAX}} = 500$ µs (dash-dotted black lines around ⋆). Multiple hexagon excitation (HEXa and HEXb) for a Hex Quasi-monopolar (HQ) stimulation pattern at 50 Hz.

### 4.3. Hex Quasi-Monopolar Multi-Site Interaction

As expected, comparing the DC offset currents (Figure 7) due to a single stimulation site (Figure 6b) with those (Figure 9) due to two stimulation sites (Figure 8), the overall DC offset current in saline increases proportionally with the number of the stimulation sites. However, in a practical case, only a few sites of the implanted electrode matrix will be activated at the same time. The stimulation sites operate simultaneously or alternately for a variable time depending on the specific application in which the neural stimulation is involved. For this reason, measurements were performed also activating separately only one of the two different hexagonal sites on the same electrode matrix with the same return electrode. This is because any possible dynamic excitation event that involves contiguous hexagonal sites is between the condition in which the two sites are simultaneously stimulated (as in

Figure 8) and the one in which they are stimulated stand-alone (SA) employing the same return electrode in the saline bath (as in Figure 11a,b). The relations among the DC offset currents in the case in Figure 11a,b are respectively:

$$i_{0aSA} = \sum_{j=1}^{6}(i_{jaSA}) + i_{REa} \tag{6}$$

$$i_{0bSA} = \sum_{j=1}^{6}(+i_{jbSA}) + i_{REb} \tag{7}$$

so that:

$$i_{0aSA} + i_{0bSA} = \sum_{j=1}^{6}(i_{jaSA} + i_{jbSA}) + i_{REa} + i_{REb} \tag{8}$$

In Figure 12, the sum of the DC current offsets of the central electrodes of the two hexagonal stimulation sites operating simultaneously $i_{0a} + i_{0b}$ (as shown in Figure 10) is compared with the sum of the DC offset currents of the same electrodes generated when each hexagonal site operates stand-alone $i_{0aSA} + i_{0bSA}$ (as represented in Figure 11a,b). The nominal injected charge $Q_{PULSE} = I_{PULSE} \cdot T_{PULSE}$ versus the DC offset current for a constant stimulation current $I_{PULSE} = I_{PULSE_{MAX}} = 460$ μA (among dashed red lines for the sum of the two central electrodes DC offset currents $i_{0a} + i_{0b}$) and for a constant pulse width $T_{PULSE} = T_{PULSE_{MAX}} = 500$ μs (among dash-dotted green lines for the sum of the two central electrodes DC offset currents $i_{0a} + i_{0b}$).

The comparison shows that the presence of the return electrode shared among two active stimulation sites, is less effective in terms of DC offset current reduction than when it operates with an only hexagonal site alternately one after the other, so that:

$$i_{0aSA} + i_{0bSA} < i_{0a} + i_{0b} \tag{9}$$

In other words, assuming an equal number of active excitation sites per area for a defined amount of time, an excitation of adjacent sites operating alternately stand-alone generates less DC offset current than a static one. This suggests how a dynamic excitation of adjacent sites is intrinsically safer than a static one as concerns the DC offset current generation.

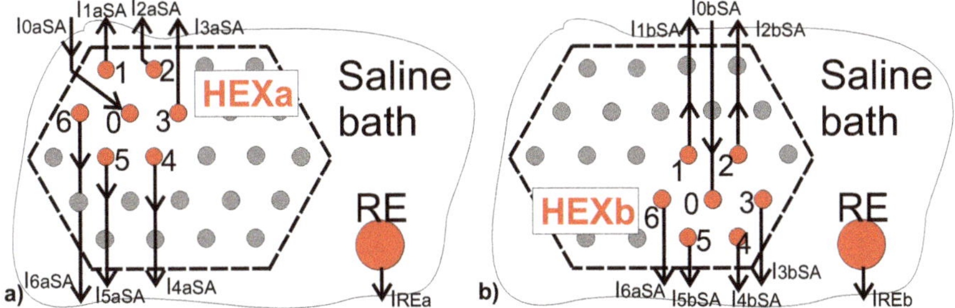


**Figure 11.** Sketch of Hex Quasi-monopolar (HQ) return stimulation patterns using (**a**) the hexagonal site HEXa operating stand-alone (SA) and the return electrode (RE) or (**b**) the hexagonal site HEXb operating SA and the RE.

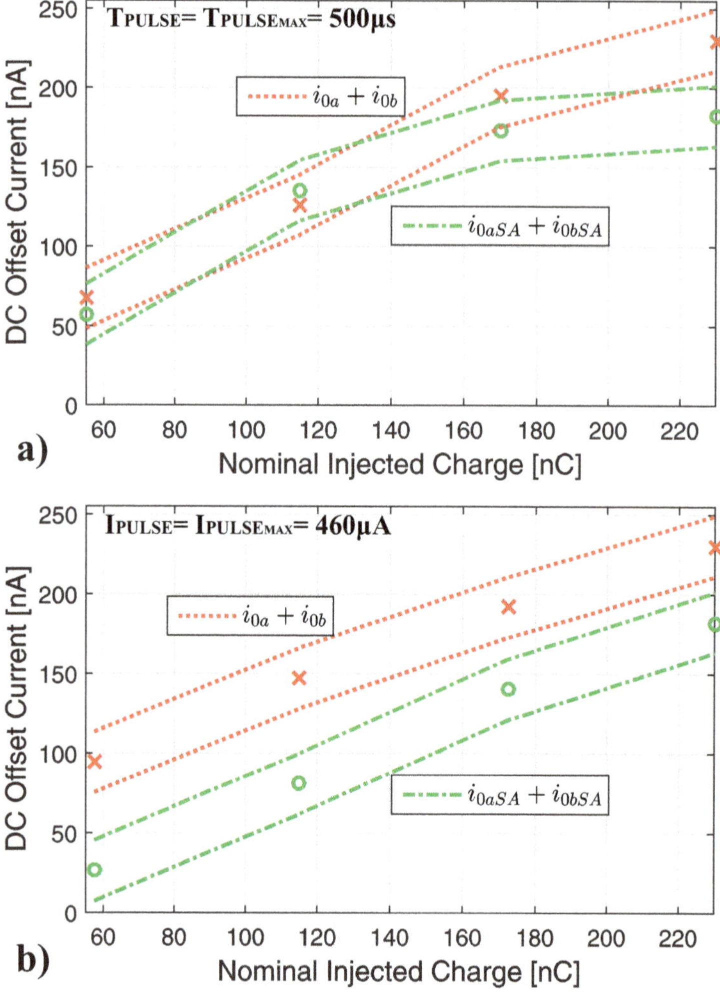



**Figure 12.** Hex Quasi-monopolar (HQ) stimulation pattern as in Figure 11a,b at 50 Hz. Comparison between the current source to the hexagonal site HEXa and HEXb excited simultaneously ($\times$: $i_{0a} +$ $i_{0b}$ in Figure 8) and the sum of the current source to a hexagonal site activated stand-alone (SA) ($\circ$: $i_{0aSA} + i_{0bSA}$ in Figure 11). (**a**) Nominal injected charge $Q_{PULSE}$ versus DC offset current ($\pm 3\sigma$ range around the average value defined by dashed and dash-dotted lines) for a pulse width constant equal to $T_{PULSE MAX} = 500$ µs. (**b**) Nominal injected charge $Q_{PULSE}$ versus DC offset current ($\pm 3\sigma$ range around the average value defined by dashed and dash-dotted lines) for a stimulation current constant equal to $I_{PULSE} = I_{PULSE MAX} = 460$ µA.

## 5. Conclusions

An automated test bench aimed at monitoring the nanoampere range DC offset current generated by platinum electrodes during biphasic stimulation in a saline bath was considered. Data storage redundancy to achieve high reliability in the measurement was implemented, and the overall procedures were described. The investigation in this paper showed how the presence of a big return electrode reduces the DC offset current. For a defined charge to be injected, DC offset current generation is related mostly to the timing of the biphasic stimulation in terms of its frequency and pulse width. In particular, for a defined injected charge needed to stimulate the tissues, lowering the frequency of the biphasic signal and a shortened pulse width reduce its DC offset current. Moreover, for the same injected charge, the sum of the DC offset currents due to the solitary activation of two hexagonal stimulation sites is always lower than the DC offset currents due to two hexagonal stimulation sites activated simultaneously. This experimental evidence suggests how the more dynamic is the stimulation pattern among the different stimulation sites with a shortened biphasic pulse width, the more safety compliant the stimulation is, as less DC offset currents are spread into the tissues.

DC offset currents depend on the voltage offset of the platinum electrodes and on the input/output impedance of the current sources (stimulator branches). Even though the stimulator is perfectly charge balanced and has an infinite input impedance, the mismatch between the macroscopic electrode-tissue interfaces generates DC offset currents in a comprehensive and realistic scenario during a biphasic stimulation. Therefore, precise verification tests as those described in this paper are needed to evaluate the effective DC offset current due to the biphasic stimulation. This is to address the electrical parameter constraints needed to guarantee the safe operating condition required by assessing the implantees' safety.

**Funding:** This work has been supported by the Department of Education and Training of the Australian Government in the framework of the 2016 Endeavour Research Fellowship granted to Dr. Aiello.

**Acknowledgments:** The author thanks Prof. Nigel Lovell, Prof. Torsten Lehmann and the Bionic Vision Australia staff of the University of New South Wales (UNSW), Sydney, Australia, for their support.

**Conflicts of Interest:** The author declares no conflict of interest.

## References

1. Lilly, J.; Hughes, J.; Alvord, E.; Galkin, T. Brief, Noninjurious Electric Waveform for Stimulation of the Brain. *Science* **1955**, *121*, 468–469. [CrossRef] [PubMed]
2. Huang, C.Q.; Carter, P.M.; Shepherd, R.K. Stimulus induced pH changes in cochlear implants: An in vitro and in vivo study. *Ann. Biomed. Eng.* **2001**, *29*, 791–802. [CrossRef] [PubMed]
3. Merrill, D.R.; Bikson, M.; Jefferys, J.G. Electrical stimulation of excitable tissue: Design of efficacious and safe protocols. *J. Neurosci. Methods* **2005**, *141*, 171–198. [CrossRef] [PubMed]
4. John, S.E.; Shivdasani, M.N.; Leuenberger, J.; Fallon, J.B.; Shepherd, R.K.; Millard, R.E.; Rathbone, G.D.; Williams, C.E. An automated system for rapid evaluation of high-density electrode arrays in neural prostheses. *J. Neural Eng.* **2011**, *8*, 036011. [CrossRef] [PubMed]
5. Brummer, S.B.; Turner, M.J. Electrical stimulation with pt electrodes. II. estimation of maximum surface redox (theoretical nongassing) limits. *IEEE Trans. Biomed. Eng.* **1977**, *BME-24*, 440–443. [CrossRef] [PubMed]
6. Aran, J.M.; Wu, Z.Y.; Cazals, Y.; de Sauvage, R.C.; Portmann, M. Electrical stimulation of the ear: Experimental studies. *Ann. Otol. Rhinol. Laryngol.* **1983**, *92*, 614–620. [CrossRef] [PubMed]
7. Hurlbert, R.J.; Tator, C.H.; Theriault, E. Dose-response study of the pathological effects of chronically applied direct current stimulation on the normal rat spinal cord. *J. Neurosurg.* **1993**, *79*, 905–916. [CrossRef] [PubMed]
8. Shepherd, R.K.; Linahan, N.; Xu, J.; Clark, G.M.; Araki, S. Chronic electrical stimulation of the auditory nerve using non-charge-balanced stimuli. *Acta Otolaryngol.* **1999**, *119*, 674–684. [CrossRef] [PubMed]
9. Xu, J.; Shepherd, R.K.; Millard, R.; Clark, G.M. Chronic electrical stimulation of the auditory nerve at high stimulus rates: A physiological and histopathological study. *Hear Res.* **1997**, *105*, 1–29. [CrossRef]
10. Franke, M.; Bhadra, N.; Bhadra, N.; Kilgore, K. Direct current contamination of kilohertz frequency alternating current waveforms. *J. Neurosci. Methods* **2014**, *232*, 74–83. [CrossRef]
11. Huang, C.Q.; Shepherd, R.K.; Carter, P.M.; Seligman, P.M.; Tabor, B. Electrical stimulation of the auditory nerve: Direct current measurement in vivo. *IEEE Trans. Biomed. Eng.* **1999**, *46*, 61–470. [CrossRef] [PubMed]
12. Carter, P.; Money, D. Feedback System to Control Electrode Voltages in a Cochlear Stimulator and the Like. U.S. Patent US5674264, 7 October 1996.
13. Huang, C.Q.; Shepherd, R.K.; Seligman, P.M.; Clark, G.M. Reduction in excitability of the auditory nerve following acute electrical stimulation at high stimulus rates. III. capacitive versus non-capacitive coupling of the stimulating electrodes. *Hear. Res.* **1998**, *116*, 55–64. [CrossRef]
14. Dommel, N.; Wong, Y.T.; Lehmann, T.; Dodds, C.W.; Lovell, N.H.; Suaning, G.J. A CMOS retinal neurostimulator capable of focused, simultaneous stimulation. *J. Neural Eng.* **2009**, *6*, 035006. [CrossRef] [PubMed]

15. Greenwald, E.; Maier, C.; Wang, Q.; Beaulieu, R.; Etienne-Cummings, R.; Cauwenberghs, G.; Thakor, N. A CMOS Current Steering Neurostimulation Array With Integrated DAC Calibration and Charge Balancing. *IEEE Trans. Biomed. Circuits Syst.* **2017**, *11*, 324–335. [CrossRef] [PubMed]

16. Luo, Z.; Ker, M. A High-Voltage-Tolerant and Precise Charge-Balanced Neuro-Stimulator in Low Voltage CMOS Process. *IEEE Trans. Biomed. Circuits Syst.* **2016**, *10*, 1087–1099. [CrossRef] [PubMed]

17. Lehmann, T.; Chun, H.; Preston, P.; Suaning, G. Current-limited passive charge recovery for implantable neuro-stimulators: Power savings, modelling and characterisation. In Proceedings of the 2010 IEEE International Symposium on Circuits and Systems, Paris, France, 30 May–2 June 2010; pp. 3128–3131.

18. Chun, H.; Yang, Y.; Lehmann, T. Safety ensuring retinal prosthesis with precise charge balance and low power consumption. *IEEE Trans. Biomed. Circuits Syst.* **2013**, *8*, 108–118. [CrossRef] [PubMed]

19. Rose, T.L.; Robblee, L.S. Electrical stimulation with Pt electrodes. VIII. Electrochemically safe charge injection limits with 0.2 ms pulses. *IEEE Trans. Biomed. Eng.* **1990**, *37*, 1118–1120. [CrossRef] [PubMed]

20. Linear Technology. Single Resistor Gain Programmable, Precision Instrumentation Amplifier. Technical Data. Available online: http://cds.linear.com/docs/en/datasheet/1167fc.pdf (accessed on 23 July 2020).

21. Analog Device. LC$^2$MOS 16-Channel High Performance Analog Multiplexers, Technical Data. Available online: http://www.analog.com/media/en/technical-documentation/data-sheets/ADG406_407_426.pdf (accessed on 23 July 2020).

22. Moghadam, G.K.; Wilke, R.; Suaning, G.J.; Lovell, N.H.; Dokos, S. Quasi-monopolar stimulation: A novel electrode design configuration for performance optimization of a retinal neuroprosthesis. *Public Libr. Sci.* **2013**, *8*, 73130. [CrossRef] [PubMed]

23. Wong, Y.T.; Dommel, N.; Preston, P.; Hallum, L.E.; Lehmann, T.; Lovell, N.H.; Suaning, G.J. Retinal Neurostimulator for a Multifocal Vision Prosthesis. *IEEE Trans. Neural Syst. Rehabil. Eng.* **2007**, *150*, 425–434. [CrossRef] [PubMed]

24. Matteucci, P.B.; Chen, S.C.; Dodds, C.; Dokos, S.; Lovell, N.H.; Suaning, G.J. Threshold analysis of a quasimonopolar stimulation paradigm in visual prosthesis. In Proceedings of the 2012 Annual International Conference of the IEEE Engineering in Medicine and Biology Society, San Diego, CA, USA, 28 August–1 September 2012; pp. 2997–3000.

 *electronics*

*Article*

# Multichannel Biphasic Muscle Stimulation System for Post Stroke Rehabilitation

Tyler Ward [1,*], Neil Grabham [1], Chris Freeman [1], Yang Wei [1,†], Ann-Marie Hughes [2], Conor Power [2], John Tudor [1] and Kai Yang [1]

1   School of Electronics and Computer Science, Faculty of Engineering and Physical Sciences, University of Southampton, Southampton SO17 1BJ, UK; njg@ecs.soton.ac.uk (N.G.); cf@ecs.soton.ac.uk (C.F.); yang.wei@ntu.ac.uk (Y.W.); mjt@ecs.soton.ac.uk (J.T.); ky2e09@soton.ac.uk (K.Y.)
2   Health Sciences, Faculty of the Environment and Life Sciences, University of Southampton, Southampton SO17 1BJ, UK; A.Hughes@soton.ac.uk (A.-M.H.); conor.power@soton.ac.uk (C.P.)
*   Correspondence: t.ward@soton.ac.uk
†   Current address: Department of Engineering at Nottingham Trent University, Nottingham NG1 4FQ, UK.

Received: 30 June 2020; Accepted: 14 July 2020; Published: 17 July 2020

**Abstract:** We present biphasic stimulator electronics developed for a wearable functional electrical stimulation system. The reported stimulator electronics consist of a twenty four channel biphasic stimulator. The stimulator circuitry is physically smaller per channel and offers a greater degree of control over stimulation parameters than existing functional electrical stimulator systems. The design achieves this by using, off the shelf multichannel high voltage switch integrated circuits combined with discrete current limiting and dc blocking circuitry for the frontend, and field programmable gate array based logic to manage pulse timing. The system has been tested on both healthy adults and those with reduced upper limb function following a stroke. Initial testing on healthy users has shown the stimulator can reliably generate specific target gestures such as palm opening or pointing with an average accuracy of better than 4 degrees across all gestures. Tests on stroke survivors produced some movement but this was limited by the mechanical movement available in those users' hands.

**Keywords:** electrode array; functional electrical stimulation; rehabilitation

## 1. Introduction

Stroke is a major cause of disability in the world. In 2017 there were approximately 12 million strokes globally [1]; prevalence is increasing due to improved longevity and rising obesity-related lifestyle factors (e.g., hypertension). Approximately 40% of stroke survivors are left with some loss of arm function as a result of their stroke [2]. Repetitive task-oriented training has been shown to help recovery as part of a rehabilitation programme [3] and one of the most prevalent technologies used to assist in rehabilitation is Functional Electrical Stimulation (FES) [4]. This method uses pulses of electrical current passing through underlying nerves, thereby causing the user's muscles to contract and generate movement. By repeated use of FES to facilitate task practice, some users may regain functional movement [5] in the affected limbs.

The SMARTmove project [6], funded by the UK Medical Research Council (MRC), realised a new FES system which provided high selectivity stimulation of the muscles used in hand, wrist and arm movements. The wearable system created by the SMARTmove project is the first example of combining a fabric FES array, a non-contact feedback sensor, and a controller able to precisely adjust stimulation using dynamic models of the arm. Although this wearable system is specifically designed for stroke survivors, who often exhibit movement dysfunction in one or more affected limbs, it can also be used in the rehabilitation of other neurological conditions (e.g., spinal cord injuries, multiple sclerosis) to restore lost or impaired movement.

Current commercial FES systems primarily use two pairs of large electrodes (typically of the order of 5 cm × 5 cm in size), giving the ability to drive two groups of muscles, but these are unable to produce selective movement such as fine finger-level gestures. A few commercial systems such as the Bioness-H200, are able to more selectively drive several groups of muscles allowing for more precise gestures; the Bioness-H200 can control five different muscles in the hand/wrist. Academic research developing transcutaneous multi-pad electrodes and arrays has been driven by the University of Belgrade [7], ETH Zurich [8], the Swiss Federal Institute of Technology Lausanne [9], Sheffield Hallam University [10], together with Fatronik-Tecnalia [11]. The layout of electrode arrays often assumes a generic pattern, and is typically used to allow the placement of a virtual electrode at any location within the array [12]. Such solutions, however, are only able to control a few of these points at a time using sequential stimulation; this limits the ability to produce complex gestures. Other systems employ a multi-pad electrode layout which is designed to match the geometry of underlying muscles, however, these require precise alignment. In both cases of electrode layout, electrodes are activated in an open-loop fashion, and do not receive real-time position feedback to correct the resulting motion.

Existing systems also have limitations in terms of the electrode construction which restricts usability. The first widely-used approach uses embroidered conductive thread resulting in a rough electrode surface which causes an uneven current distribution, leading to discomfort when the stimulation current is high. In addition, embroidered high density conductive areas such as electrodes are stiff. The second widely-used approach uses gel electrodes on plastic substrates which are not sufficiently flexible, breathable, or comfortable.

The SMARTmove project has created a new FES system to address the limitations of existing FES techniques. By using an array of small electrodes, rather than a pair of large electrodes, a high level of selectivity in the stimulated muscles has been achieved. Independent control of smaller electrodes also enables more individual muscles to be activated simultaneously, thus increasing the range of achieved gestures. By using real-time feedback and an iterative learning control approach, the system can help the user achieve precise movements while adapting to their physiology, rather than being limited to a set of fixed stimulation patterns. This allows the stimulation to be better tailored to the user's progress and condition and avoids the need for exact electrode positioning. The printed dry fabric electrode array also makes the system easier to don/doff and avoids issues that arise with gel electrodes, such as drying out, susceptibility to contamination, and a short life time (typically 1 to 2 weeks). The fabric electrode array is more flexible and comfortable. It is also washable and reusable, and therefore cost effective for long term use.

This paper focuses on the design of the stimulation electronics used in the SMARTmove system. A review of stimulation electronics output stages for FES has been carried out by Souza [13]. The applied stimulation can be grouped into two general categories: monophasic where the stimulation pulses only have one polarity; and biphasic where the stimulation pulses have both positive and negative polarities. The stimulator design reported in this paper uses biphasic stimulation as this reduces the potential for skin burns and tissue damage [14] and provides a wider range of stimulation profiles when compared to monophasic stimulation. Of the biphasic designs investigated, the majority use H-bridge drivers for controlling the phase of the stimulation. However, this requires a pair of electrodes for each stimulation channel. Other non H-bridge designs often also require control of both the source and return of each channel, such as that by Huerta [15]. Where a common return electrode is used, as is the case when using an array, each channel needs to be able to generate both positive and negative stimulation voltages rather than swapping the polarity of the electrodes, as typically occurs with simple electrode pairs.

Existing FES stimulator designs are also often based around discrete components rather than using an integrated package, thereby increasing the circuit board space required for each channel.

Many existing designs also use an isolation transformer on the output to provide an isolation barrier; these transformers often have a significant footprint which leads to a large implemented size when there are multiple channels.

Using existing solutions for FES stimulation would result in a bulky and complex stimulator that would be unsuitable for wearable use. To overcome these limitations the developed design uses off the shelf multichannel high voltage switchIntegrated circuits (ICs) combined with discrete current limiting and dc blocking circuitry for the stimulation frontend. Pulse timing is controlled by Field Programmable Gate Array (FPGA) based logic which is in turn controlled by a microprocessor to manage communication with the software managing the stimulation profiles. These provide greater parameter control than previous designs with a low cost and a small size suitable for wearable deployment.

In this paper we discuss, in Section 2, the constituent parts and operation of the rehabilitation solution as a whole. The designed electronics that make up the multichannel biphasic stimulator is discussed in Section 3. Section 4 presents system use and testing on both healthy individuals and stroke survivors, illustrating use of the system to stimulate individual finger movements to produce a range of gestures. Finally, conclusions are provided.

## 2. Complete System: Components and Operation

The SMARTmove system is designed to allow stroke survivors to practice a series of commonly undertaken tasks at home with the assistance of electrical stimulation, as part of their rehabilitation programme. Currently the system supports three different gestures (see Figure 1): palm flattening, pointing and pinching. These gestures form the basis for functional activities such as closing a drawer or operating a light switch.

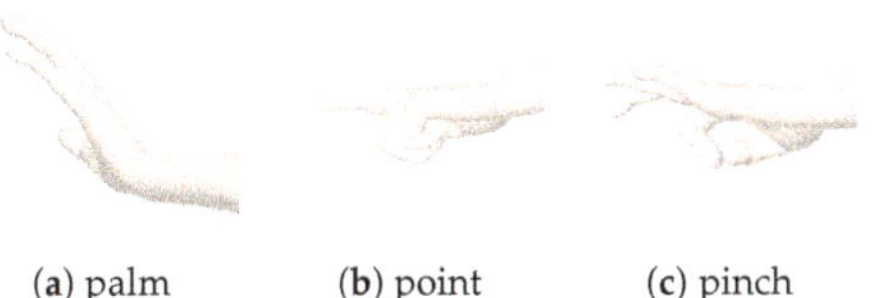

(a) palm      (b) point      (c) pinch

**Figure 1.** Target gestures the system will assist the user in creating.

The SMARTmove system contains several components that work together to control and deliver the electrical stimulation. A graphical overview of the system components can be seen in Figure 2.

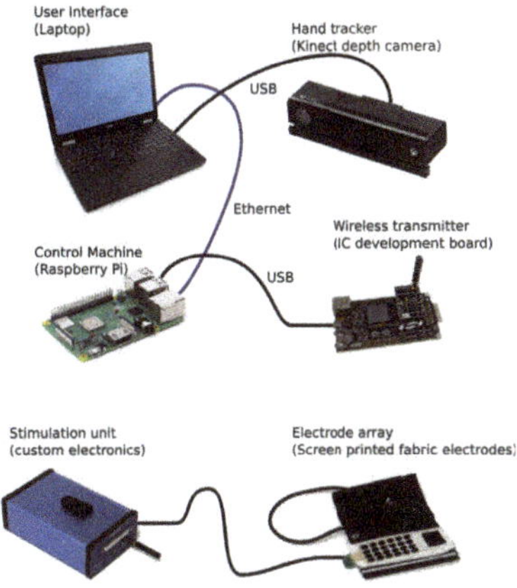


**Figure 2.** Overview of the SMARTmove system.

The SMARTmove system uses a commercial, off the shelf, depth camera (Microsoft Kinect v2) to capture the movements of a user's hand. The depth information from the camera is processed on a laptop to extract the angular position of the user's hand and fingers. This processing is performed by a standalone application which then streams the data into the Graphical User Interface (GUI) application also running on the user's laptop.

The GUI provides the interface for the user to control the system. After initial configuration of the system a series of tasks of increasing complexity are presented to the user. The system uses a gamification approach to the rehabilitation process by enabling users to progress from task to task once they have gained enough points in a given task. These points are based on how well they perform a task using performance metrics, such as the accuracy of the final gesture and the smoothness of movement. This means that the user will progress onto the later tasks more quickly once they have become more proficient in the earlier tasks. The GUI application translates the user's interactions into a series of commands for the control software and displays the returned feedback information to the user after every task.

The control system [16] runs on a separate Raspberry Pi low-cost computer and receives the commands and hand tracking data from the GUI on the laptop. This is achieved using a network connection over a wired Ethernet link. The software has a map of which electrode causes which group of hand joint angles to move, and by how much. This map is automatically generated by monitoring the movement during an array identification sequence (described further in Section 4.1). An iterative learning controller is used in the control system, allowing it to improve the user's response by adjusting the delivered stimulation based on previous results. The stimulation levels which are calculated by the controller forty times a second are then sent over USB to the wireless transmitter which relays them wirelessly to the stimulator unit electronics (described in detail in Section 3).

The stimulation pulses are delivered to the user through a fabric electrode array [17], connected via a cable to the stimulation unit. This array consists of screen printed dry electrodes in a six by four pattern, giving twenty four stimulation electrodes in total, together with an additional large common return electrode. The electrode array is attached to a cuff which is placed on the arm and secured using a hook and loop fastener. The cuff applies pressure to the electrode array ensuring good contact between the electrodes and the user's skin.

## 3. Stimulator Unit: Electronics Design and Implementation

### 3.1. Overview

The stimulator unit converts the stimulation levels sent from the control software into the electrical stimulation pulses. Each set of levels from the controller triggers a single set of synchronised pulses from the unit. Each of the twenty four independent stimulation channels can have a different stimulation intensity level, including the possibility of zero stimulation. The stimulator uses voltage controlled stimulation in a multichannel configuration with varying pulse length used to adjust the stimulation intensity of each channel. Each channel has an independently controllable pulse width and waveform, adjustable on every pulse. The stimulation voltage is common across all the electrodes but is variable between 50 V and 100 V DC.

The design is implemented using a series of interconnected modules split across several custom printed circuit boards (PCBs). The modules are separated into a control board together with sufficient eight channel driver boards to produce the required number of stimulation channels. The setup for the SMARTmove system uses three driver boards to achieve twenty four channels. These boards are slotted inside a compact enclosure, shown in Figure 3, which can be worn, for example, on the user's belt. The boards are connected together with a series of short cable harnesses. The stimulator is connected to the fabric electrode array worn on the arm using a detachable multicore cable. Figure 4 is a schematic showing the functionality implemented on each board.

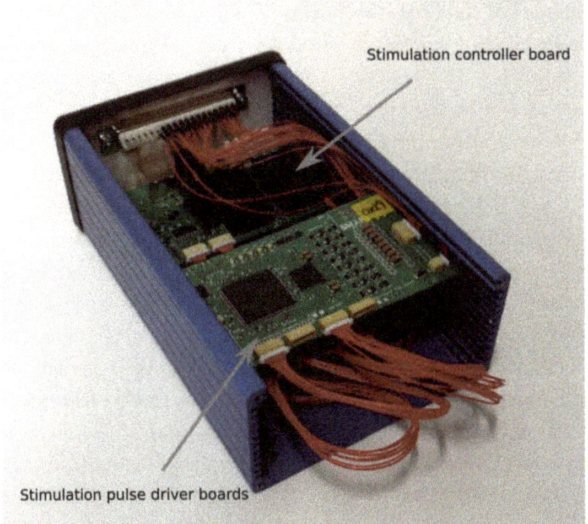


**Figure 3.** Interior of custom stimulation unit.

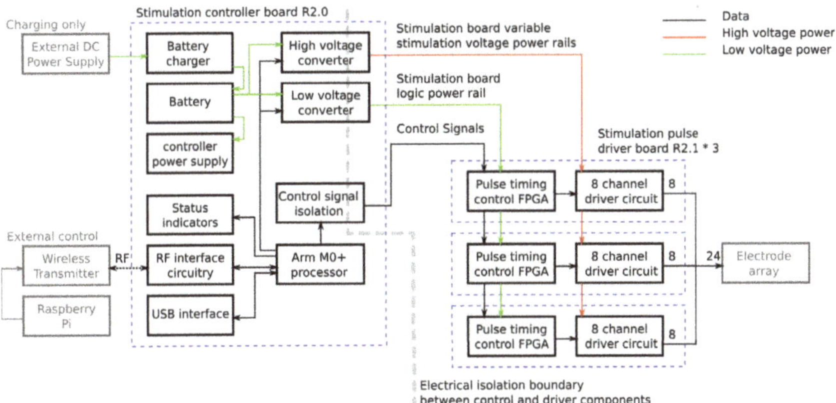


**Figure 4.** Block diagram of stimulator electronics.

## 3.2. Stimulation Controller Board

The controller board contains the modules for managing the system and passing commands from the control program to the driver boards. Commands are received from the control software running on the Raspberry Pi either over a USB cable or via a wireless link from a USB attached transmitter. Commands are processed by an ARM M0+ microprocessor which has inbuilt USB and wireless connectivity. Each stimulation request from the control program defines a single set of pulses, one for each channel, with forty stimulation requests being sent per second. Each set of pulses is defined separately which allows the length of every pulse to be varied, and prevents stimulation continuing should the connection to the Raspberry Pi be lost. When a message is received it is checked for corruption and, if it is valid, it is decoded into the pulse lengths for each channel and the voltages for the stimulation power supplies. The high voltage power supplies are reconfigured to provide a new stimulation voltage if it has changed. The pulse lengths for each channel are then sent to the corresponding driver over an $I^2C$ interface. In order to keep the pulses synchronised, a dedicated start signal is used which triggers all of the drivers to start the pulses at the same time.

Another major function of the control board is management of the power supplies. To provide for wireless operation an internal battery is used to power the stimulator and all other required voltages are generated from this. The battery, which provides approximately four hours of active use, is charged using an external 9 V DC external power supply which can also be used to power the unit. Internal power regulators take either the battery voltage or DC input voltage and generate 3.3 V DC for the microprocessor and a switchable 5 V DC supply for powering the rest of the system.

The stimulation voltage is generated using a pair of isolated variable DC-DC voltage converters which take the power from the 5V power rail and produce between 50 V and 100 V DC which is selectable by the controller for each pulse. The DC-DC converters sit across an isolation boundary which separates the low voltage interface circuitry on the board from the high voltage stimulation circuitry located on the other boards. This boundary adds two safety features. In the event of major damage to the driver boards the high stimulation voltages cannot appear on any of the interface connectors. In addition, current cannot flow between the electrode interface and any of the other connectors on the unit. This means the stimulation current will only flow between the stimulation electrodes and the return electrode and will not take any unwanted paths through the user to an interface connector if touched. Power for the low voltage circuitry on the driver boards is generated by another isolated DC-DC converter at 5 V DC. Control signals for the driver boards are passed through optical isolation ICs in order to maintain the isolation barrier.

### 3.3. Stimulation Pulse Driver Board

The stimulation pulse's driver boards receive pulse length settings and triggering from the control board and produce the required pulses. Each pulse is made up of a variable length positive pulse followed by a variable length negative pulse. The electrode is then actively driven to 0 V after each pulse, where it remains until the next pulse.

The pulse drive circuitry, which generates the initial pulse waveform, consists of a constant voltage pulse generator which is capable of operating at high voltages. The pulse output is passed through a current limiting circuit and a DC blocking capacitor. The connections between the components can be seen in Figure 5.

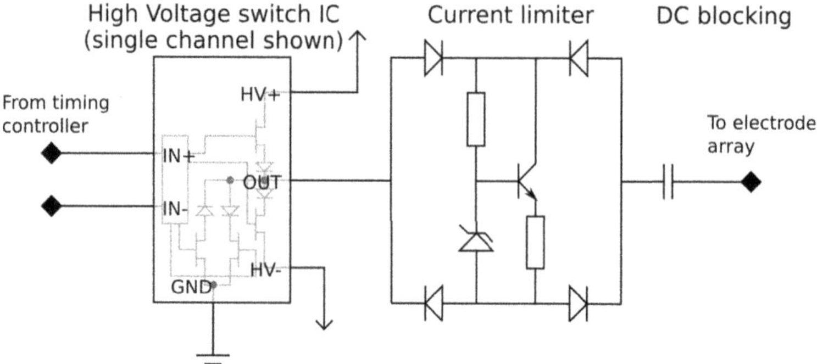


**Figure 5.** Circuit schematic of stimulation pulse driver board output stage.

The pulse generator uses a commercially available high voltage switch integrated circuit (IC). This IC contains a Field Effect Transistor (FET) based switch stage along with the circuitry to control the gates from logic level input signals. Each of the eight outputs can be driven to one of three states (+V, Gnd, −V) in any combination. This creates the positive and negative pulses as well as an active return to ground when not producing a pulse. Using an off the shelf IC provides a very compact implementation which fits in a 10 mm by 10 mm package.

In order to achieve user safety, protective circuitry is added to the output of each channel. Current limiting circuitry is placed after every output to prevent large stimulation currents from flowing if there is a low resistance path through the user. A Zener diode based current limiter is set to a 9 mA limit and placed within a diode bridge to limit the current in either direction. The final component in the pulse path is a DC blocking capacitor which prevents any DC voltage appearing on the electrodes which could potentially cause electrical burns in the event of hardware failure. These nine extra components per channel require minimal board space, each taking 80 mm$^2$ of PCB surface area per channel. Once the driver IC and routing space to link the parts together are taken into account the required area is still very low; the eight channel driver along with the required extra circuitry will fit in an area of 20 mm $\times$ 35 mm assuming a 4-layer PCB, of which 16 mm $\times$ 20 mm comprises the extra safety circuitry. The integrated driver design also reduces the cost of the stimulator compared with using discrete control transistors for each channel.

The pulse generator is driven by a timing controller which controls the pulse lengths on each channel. The timing controller is implemented on a Field Programmable Gate Array (FPGA) which contains an output control block for each output and a communication module to provide an interface to the controller board. In the developed design each driver IC has its own control FPGA in order to allow the number of channels to be easily altered if required. In a less flexible hardware configuration a control FPGA could control multiple driver ICs to reduce cost/size. The output control block contains a high accuracy timer and state machine which cycles through the required stages of the pulse. The control blocks in the FPGA have several waveforms available allowing them to produce a range of output pulse waveforms as can be seen in Figure 6. The default mode used in the majority of the tests to date was the biphasic negative mode which uses a matched pair of pulses where the negative pulse is immediately followed by a positive pulse as shown in Figure 7.

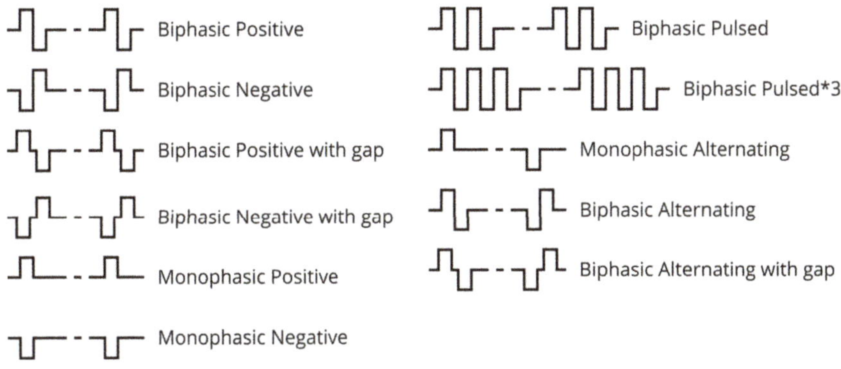


**Figure 6.** Stimulator output waveform options.

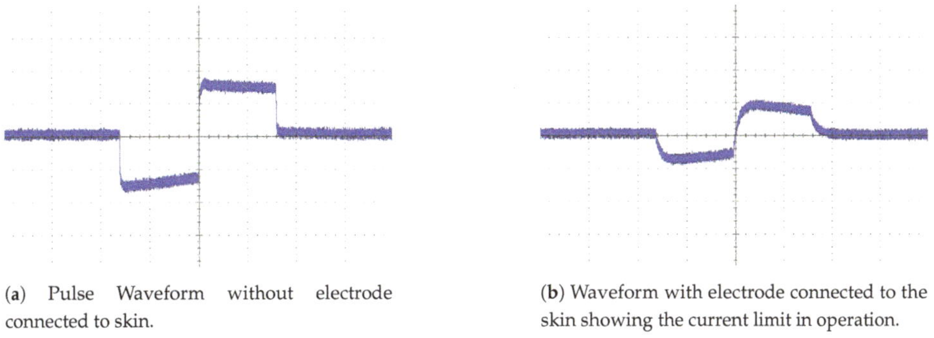

(a) Pulse Waveform without electrode connected to skin.

(b) Waveform with electrode connected to the skin showing the current limit in operation.

**Figure 7.** Stimulator pulse waveforms (oscilloscope settings 50 V/div, 100 µs/div).

The controller board interface is in two parts. The pulse lengths are sent to the drivers using the isolated I$^2$C interface on the control board. The controller addresses each of the driver boards in turn sending them the settings for each channel for the next pulse. The start trigger signal is then sent using a dedicated logic signal to all boards at once. This allows all the boards to trigger the output control blocks thus starting their outputs close to simultaneously. Differences in timing of up to 20 ns between the outputs of different driver boards are possible as the FPGA clocks are not synchronised but this is negligible compared to the typical pulse lengths of 60 µs to 500 µs. A simplified block diagram showing how these interface with the rest of the components in the timing controller can be found in Figure 8.

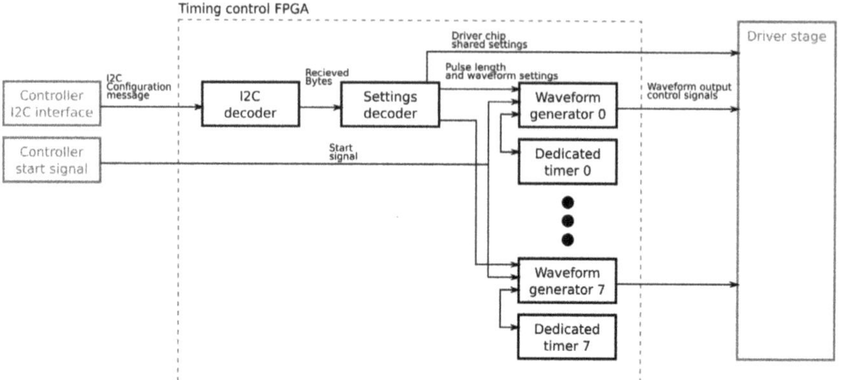


**Figure 8.** Simplified block diagram of Field Programmable Gate Array (FPGA) logic blocks used to generate pulse waveforms.

## 4. System Testing

### 4.1. System Setup

The cuff containing the electrode array is first placed on the arm and secured in place. Exact placement is not required as the control system will automatically learn and adjust the stimulation according to measured performance. The next stage sets the maximum stimulation level. This needs to be undertaken before first use, each time the system is used, as conditions such as humidity and skin hydration level can affect a user's response on any given day. Generally, higher stimulation levels result in more precise movements, however, with high stimulation levels discomfort can occur. The stimulation level is adjusted to the maximum level that can be tolerated without any discomfort for the user.

The next stage of the setup process is to perform an array identification sequence which applies stimulation to each electrode in turn and the system measures the resultant movement as the stimulation level is ramped up to the set maximum level. These measured movements are stored so that they can be used to calculate the stimulation required for each target gesture. This step need only be carried out during the first few instances of using the system. During these first few uses the control system will have recorded how the user responds under different conditions or differences in array placement. Then, in future use, the control system will observe the first few gestures and use recorded data to determine how to apply the stimulation.

The hand positions for each of the three target gestures and an initial hand position are then recorded. Healthy subjects create the target gestures themselves; those with impaired movement need another person to make the reference gesture for them.

*4.2. System Use*

Once the setup is complete the system can stimulate the muscles to produce the required gestures. A target gesture is selected on the GUI application and the test is started. The system then calculates the required stimulation pattern to move from the initial position to the desired gesture, as can be seen in Figure 9. The stimulation is then delivered and the resultant motion is observed and fed back into the iterative learning control stimulation algorithm to improve the resulting gesture at the next stimulation run. The user will perform several stimulation runs for a gesture before advancing to the next gesture.

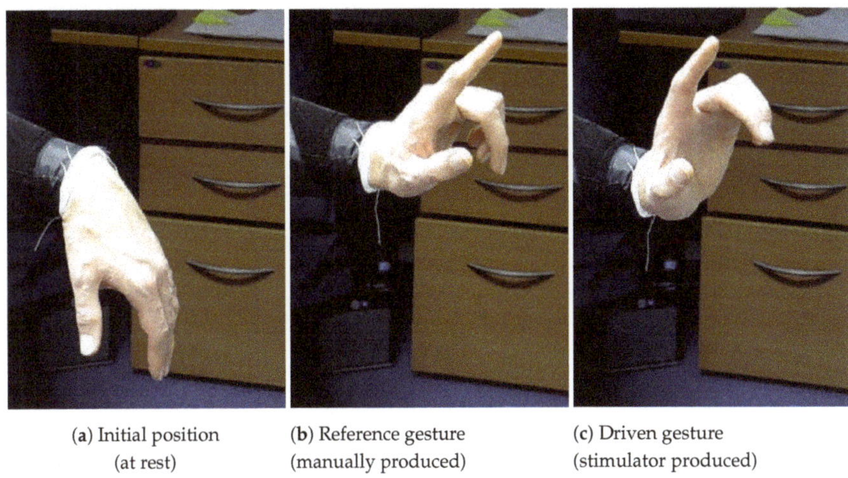

|  |  |  |
| :---: | :---: | :---: |
| (**a**) Initial position | (**b**) Reference gesture | (**c**) Driven gesture |
| (at rest) | (manually produced) | (stimulator produced) |

**Figure 9.** Photographs showing target and produced gestures.

*4.3. Testing on Healthy Volunteers and Stroke Survivors*

The stimulation electronics has been tested as part of the SMARTmove system on five healthy participants as well as three members of the End User Group (EUG) of stroke survivors.

4.3.1. Initial Tests with Healthy Participants

Initial tests were carried out on five healthy participants, following ethical approval. The required maximum stimulation level which was comfortable was found to differ significantly between users, with acceptable pulse lengths varying from 60 µs to 500 µs. The system was able to produce a range of movements when testing individual electrodes, and the gesture tests were able to produce accurate gestures. The accuracy of the driven gestures was measured by the control software using angles received from the camera tracking data. The data was processed using three functions to produce values for comparison. These were also used to provide feedback to the user through the GUI.

Equation (1) was used to calculate the 'percent performance': this is a percentage score of the movement as a proportion of the total movement required to reach the target gesture. Equation (2) provided the 'mean error': this is the average absolute error between the produced gesture and the target gesture in degrees. Equation (3) provided the 'percent ROM performance': this is the performance relative to the range of movement (ROM), a high score here shows the muscles are being stimulated to move in the correct direction even if they do not achieve the necessary scale of movement required to fully produce the gesture. In these equations $\theta$ represents the vector of joint angles after stimulation, $\theta_{REF}$ the vector of target angles, $\theta_{ROM}$ the joint angle ROM values, and $\theta_{INI}$ the initial joint angles.

$$100(1 - \frac{\|\theta_{REF} - \theta\|}{\|\theta_{REF} - \theta_{INI}\|}) \tag{1}$$

$$\|\theta_{REF} - \theta\| \tag{2}$$

$$100(1 - \frac{\|\theta_{REF} - \theta\|}{\|\theta_{REF} - \theta_{ROM}\|}). \tag{3}$$

Figure 10 shows a series of stimulation runs performed on one participant targeting the three different gestures shown in Figure 1. The mean error of the joint angles between the achieved position and the target position is shown in the centre. The left graph is the accuracy as a percentage of the total movement required, and the right graph shows the accuracy as a percentage of the range of movement.

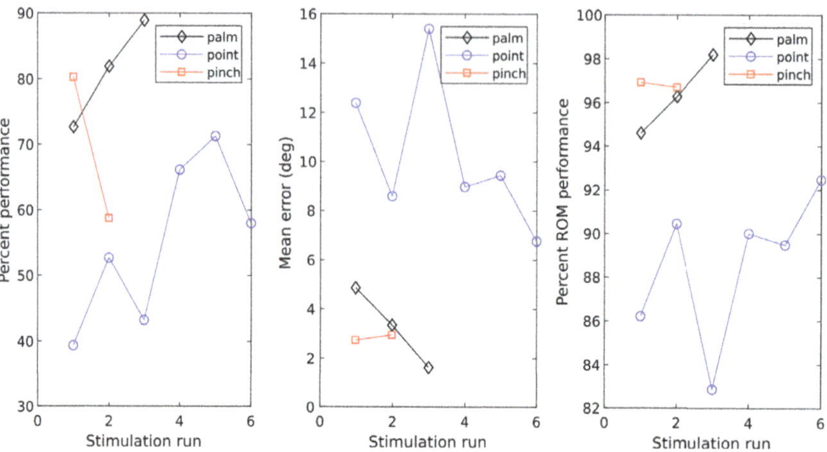

**Figure 10.** Graph showing reduction in error by iterative improvement of stimulation profile for the three different gestures.

The iterative learning algorithm altered the stimulation in subsequent stimulation runs which is reflected in the reduction in error in the centre graph as the number of stimulation runs increases. Figure 11 shows the mean error and percent performance recorded from the last run of each gesture on the five healthy participants using the same equations as in Figure 10. The performance exceeded 91% for all gestures and participants.

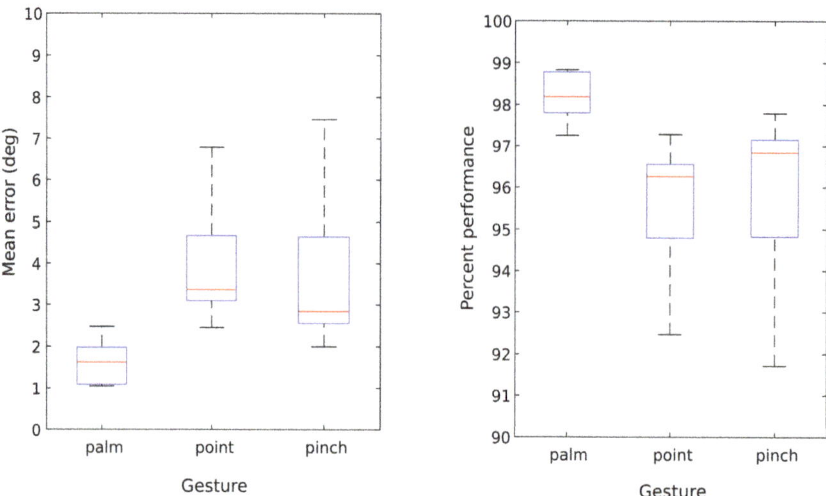

**Figure 11.** Graph comparing gesture performance on final run across participants.

### 4.3.2. Tests of Different Waveforms with Healthy Participants

Tests with different waveform patterns were carried out on the healthy participants. Across the test group participants found that changing the stimulation waveform changed the amount of movement produced. Different participants, however, found different waveforms produced stronger results in their case. For example, one participant found the biphasic negative waveform produced the strongest results, for another the biphasic alternating waveform produced stronger results. This indicates that there are benefits to selecting the waveform on a per user basis. Further research will be required to investigate the best use of differing waveform patterns.

### 4.3.3. Tests of Additional Upper Arm Stimulation with Healthy Participants

The stimulation unit has also been tested with additional electrodes allowing stimulation of the upper arm to produce elbow extension movement. The new electrode was produced using the same dry electrode materials as the array, but has two large electrode pads each 5 cm × 5 cm, one for stimulation and one as an additional return. To connect this, one of the lower arm array electrodes was disconnected and its stimulation channel used for the new upper arm electrode. The same stimulation pulses were used for the upper arm electrode as was used on the electrodes on the wrist array. With this configuration, the system was able to produce contraction in the triceps muscles in addition to stimulating extension movements in the wrist and fingers. The use of additional electrodes on the upper arm therefore offers the option to create more complex actions using both hand and arm movements.

### 4.3.4. Tests with Stroke Survivors

Initial testing of the design on the wrist (i.e., lower arm only) with two of the EUG participants produced reduced movement when compared to identical tests with healthy users. The array identification phase of the test only produced limited movement, which resulted in very poor gesture stimulation.

There was almost no sensation before the unit reached its software defined pulse length limit indicating that the maximum stimulation level was too low. Tests varying the stimulation voltage used caused no obvious difference to either the comfort or movement generated by the system. To overcome this, the software was altered to increase the maximum pulse length to 1000 μs from the previously used level of 500 μs.

Another set of tests with the increased pulse length were carried out with two EUG members, one of whom was also in the previous group. These tests showed increased movement when compared with the previous tests. The first participant had a good level of movement with the system but wrist and finger extension was limited either by increased spasticity or by possible changes in reciprocal inhibition, whereby muscles on one side of a joint relax to accommodate contraction on the other side of the joint. After several tests the idle position of the wrist and fingers was more extended and relaxed. Due to this, there was a smaller range of movement observed from the stimulation as the starting point was closer to the maximum extension achievable. The idle position of the wrist and fingers returned back to a more flexed posture after a short period of time. The other participant had more movement than in their previous tests, however, additional movement would be required to closely achieve the target gestures. Despite this the wrist and fingers were again in a more relaxed position after the tests than before using the system. This needs further investigation and further tests such as using different stimulation waveforms are planned to optimize movement for the stroke survivors.

## 5. Conclusions

The new design for a multichannel biphasic stimulator discussed in this paper unlocks the possibility of achieving compact and low cost multichannel stimulators which are suitable for wearable deployment. In addition, tests show the device to be capable of generating the desired specificity and control over stimulation allowing complex multiple muscle gestures to be produced.

The use of an integrated high voltage driver simplifies and reduces the size of the electronics compared to designs using discrete switching components. Only nine additional components are required per channel using only 80 mm$^2$ of PCB surface area in total. This means the additional circuitry for an eight channel driver fits in a 16 mm by 20 mm section of double sided PCB. In total, once the driver IC and routing space to link the parts together are taken into account an entire eight channel driver stage can be implemented in an area of 20 mm by 35 mm of 4-layer PCB space. The low cost per channel of the drivers makes large channel count stimulators feasible in terms of component and material cost.

When tested on healthy users the stimulator design, in conjunction with the rest of the SMARTmove system, is capable of creating predictable, controlled, and selective movements. Tests on stroke survivors have produced mixed results, some subjects have been limited by either spasticity in the flexor muscles, or possibly changes in the reciprocal inhibition patterns, whereas others have shown less response to the stimulator and are expected to need a stronger level of stimulation or a different waveform to achieve sufficient movement to realise the target gestures. Refinements to the system are being investigated with the aim of increasing the movement in those who had a weaker response to the stimulation provided.

**Author Contributions:** Conceptualization, C.F., J.T. and K.Y.; methodology, T.W., C.F., A.-M.H., J.T. and K.Y.; software, T.W., C.F.; validation, C.F.; formal analysis, C.F.; investigation, T.W., A.-M.H. and C.P.; resources, N.G., Y.W. and K.Y.; data curation, C.F.; writing—original draft preparation, T.W.; writing—review and editing, T.W., J.T. and K.Y.; visualization, C.F.; supervision, K.Y.; project administration, K.Y.; funding acquisition, C.F., J.T. and K.Y. All authors have read and agreed to the published version of the manuscript.

**Funding:** This research was funded by Medical Research Council (grant ref: MR/N027841/1) and EPSRC (grant ref: EP/S001654/1).

**Acknowledgments:** We would like to acknowledge the support of the SMART-move project End User Group participants and Advisory Board Members. Ethical approval was received from the Faculty of Health Sciences Ethics Committee, University of Southampton, reference 23824.A7.



## References

1. Institute for Metrics and Health Evaluation. Global Burden of Disease Tool. 2017. Available online: http://ghdx.healthdata.org/gbd-results-tool (accessed on 25 May 2020).
2. Royal College of Physicians. National Clinical Guideline for Stroke. Available online: https://www.strokeaudit.org/SupportFiles/Documents/Guidelines/2016-National-Clinical-Guideline-for-Stroke-5t-(1).aspx (accessed on 25 May 2020).
3. Pollock, A.; Baer, G.; Campbell, P.; Choo, P.L.; Forster, A.; Morris, J.; Pomeroy, V.M.; Langhorne, P. Physical rehabilitation approaches for the recovery of function and mobility following stroke. *Cochrane Database Syst. Rev.* **2014**, *9*, 965–967. [CrossRef] [PubMed]
4. Vafadar, A.K.; Côté, J.N.; Archambault, P.S. Effectiveness of functional electrical stimulation in improving clinical outcomes in the upper arm following stroke: A systematic review and meta-analysis. *BioMed Res. Int.* **2015**, *2015*, 1–14. [CrossRef] [PubMed]
5. Teasell, R.W.; Foley, N.C.; Bhogal, S.K.; Speechley, M.R. An evidence-based review of stroke rehabilitation. *Top. Stroke Rehabil.* **2003**, *10*, 29–58. [CrossRef] [PubMed]
6. SMART Move Website. Available online: www.smartmove.soton.ac.uk/ (accessed on 20 February 2020).
7. Popović-Maneski, L.; Topalović, I. EMG Map for Designing the Electrode Shape for Functional Electrical Therapy of Upper Extremities. In *Converging Clinical and Engineering Research on Neurorehabilitation III, Proceedings of the 4th International Conference on NeuroRehabilitation (ICNR2018), Pisa, Italy, 16–20 October 2018*; Springer: Berlin/Heidelberg, Germany, 2018; pp. 1003–1007. [CrossRef]

8. Kuhn, A.; Keller, T.; Micera, S.; Morari, M. Array electrode design for transcutaneous electrical stimulation: A simulation study. *Med. Eng. Phys.* **2009**, *31*, 945–951. [CrossRef] [PubMed]

9. Crema, A.; Furfaro, I.; Raschellà, F.; Micera, S. Development of a hand neuroprosthesis for grasp rehabilitation after stroke: State of art and perspectives. In *Converging Clinical and Engineering Research on Neurorehabilitation III, Proceedings of the 4th International Conference on NeuroRehabilitation (ICNR2018), Pisa, Italy, 16–20 October 2018*; Springer: Berlin/Heidelberg, Germany, 2018; pp. 89–93. [CrossRef]

10. Kenney, L.P.; Heller, B.W.; Barker, A.T.; Reeves, M.L.; Healey, J.; Good, T.R.; Cooper, G.; Sha, N.; Prenton, S.; Liu, A.; et al. A review of the design and clinical evaluation of the ShefStim array-based functional electrical stimulation system. *Med. Eng. Phys.* **2016**, *38*, 1159–1165. [CrossRef] [PubMed]

11. Velik, R.; Malešević, N.; Maneski, L.; Hoffmann, U.; Keller, T. INTFES: A Multi-pad Electrode System for Selective Transcutaneous Electrical Muscle Stimulation. In Proceedings of the 16th Annual International Functional Electrical Stimulation Society Conference, São Paulo, Brazil, 8–11 September 2011.

12. Heller, B.W.; Clarke, A.J.; Good, T.R.; Healey, T.J.; Nair, S.; Pratt, E.J.; Reeves, M.L.; van der Meulen, J.M.; Barker, A.T. Automated setup of functional electrical stimulation for drop foot using a novel 64 channel prototype stimulator and electrode array: Results from a gait-lab based study. *Med. Eng. Phys.* **2013**, *35*, 74–81. [CrossRef] [PubMed]

13. Souza, D.C.D.; Gaiotto, M.D.C.; Nogueira Neto, G.N.; Castro, M.C.F.D.; Nohama, P. Power amplifier circuits for functional electrical stimulation systems. *Res. Biomed. Eng.* **2017**, *33*, 144–155. [CrossRef]

14. Popovic, M.; Curt, A.; Keller, T.; Dietz, V. Functional electrical stimulation for grasping and walking: Indications and limitations. *Spinal Cord* **2001**, *39*, 403–412. [CrossRef] [PubMed]

15. Huerta, S.C.; Tarulli, M.; Prodic, A.; Popovic, M.R.; Lehn, P.W. A universal functional electrical stimulator based on merged flyback-SC circuit. In Proceedings of the 2012 15th International Power Electronics and Motion Control Conference (EPE/PEMC), Novi Sad, Serbia, 4–6 September 2012; pp. LS5a.3-1–LS5a.3-5. [CrossRef]

16. Freeman, C.; Spraggs, M.W.; Hughes, A.M.; Yang, K.; Tudor, M.; Grabham, N. Multiple model adaptive ILC for human movement assistance. In Proceedings of the IFAC European Control Conference 2018, Limassol, Cyprus, 12–15 June 2018; pp. 1–6. [CrossRef]

17. Yang, K.; Meadmore, K.; Freeman, C.; Grabham, N.; Hughes, A.M.; Wei, Y.; Torah, R.; Glanc-Gostkiewicz, M.; Beeby, S.; Tudor, M. Wearable electronic sleeve for muscle stimulation. In Proceedings of the Annual Conference of the International Functional Electrical Stimulation Society 2018, Nottwil, Switzerland, 28–31 August 2018.

 *electronics*

*Article*

# High-Performance Analog Front-End (AFE) for EOG Systems

Alberto López [1], Francisco Ferrero [1,*], José Ramón Villar [2] and Octavian Postolache [3,4]

[1]   Department of Electrical and Electronic Engineering, University of Oviedo, 33204 Gijón, Spain; uo181549@uniovi.es
[2]   Department of Computer Science, University of Oviedo, 33204 Gijón, Spain; villarjose@uniovi.es
[3]   Instituto de Telecomunicações, ISCTE-IUL, Av. RoviscoPais, 1, 1049-001 Lisboa, Portugal; opostolache@lx.it.pt
[4]   ISCTE-Instituto Universitario de Lisboa, Av. das Forças Armadas, 1649-026 Lisboa, Portugal
*   Correspondence: ferrero@uniovi.es; Tel.: +34-985-1825-52

Received: 18 May 2020; Accepted: 8 June 2020; Published: 11 June 2020

**Abstract:** Electrooculography is a technique for measuring the corneo-retinal standing potential of the human eye. The resulting signal is called the electrooculogram (EOG). The primary applications are in ophthalmological diagnosis and in recording eye movements to develop simple human–machine interfaces (HCI). The electronic circuits for EOG signal conditioning are well known in the field of electronic instrumentation; however, the specific characteristics of the EOG signal make a careful electronic design necessary. This work is devoted to presenting the most important issues related to the design of an EOG analog front-end (AFE). In this respect, it is essential to analyze the possible sources of noise, interference, and motion artifacts and how to minimize their effects. Considering these issues, the complete design of an AFE for EOG systems is reported in this work.

**Keywords:** analog front-end (AFE); electrooculogram (EOG); electrooculography; interference; noise; signal acquisition

---

## 1. Introduction

The biopotentials generated by the human body have given rise to numerous studies and to some applications. The biopotentials recorded by the movement of the eyes are called an electrooculogram. The origin of this recording goes back to the year 1848 when the German physicist Emil du Bois-Reymond observed for the first time in history that the front of the eyeball (the cornea) is electrically positive with respect to the back (the retina), thus concluding that the eye could be modeled as a dipole. The eye, by performing different movements within its orbital cavity, generates a measurable potential by means of conveniently arranged electrodes. This is the basis of electrooculography upon which the subsequent studies are based. Figure 1 represents in a simplified way the foundation of the EOG by modeling the eye as an electrode-measurable dipole [1].

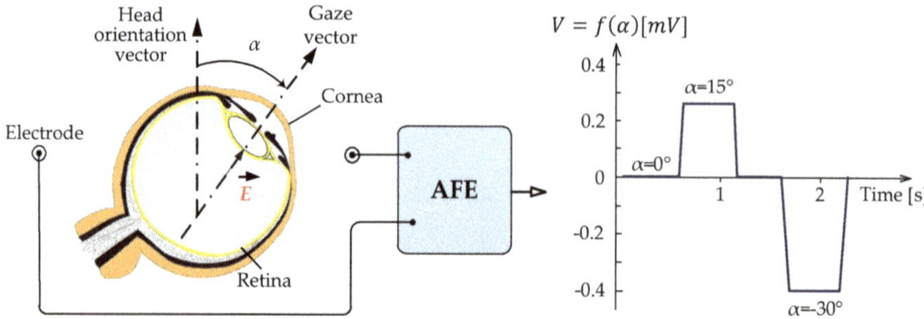


**Figure 1.** Ideal representation of the foundation of the electrooculogram (EOG) when modeling the eye as a measurable dipole using surface electrodes. These electrodes make up the input to the Analog Front-End (AFE) to record the potential difference generated by eye movement.

The recording of this biopotential allows the detection of different factors such as ocular movement or light stimulation. Besides, this recording complements other systems of diagnosis and the detection of biopotentials such as electroencephalography (EEG) or electromyography (EMG). EOG recording is routinely applied diagnostically to investigate the human oculomotor system, for instance, in sleep studies [2,3], to prevent computer vision syndrome [4], or for Ataxia SCA-2 diagnosis [5,6]. This is since eye movements provide critical signs of neurological disorders [7]. Eye movement research is also of great interest in the control of human prosthesis [8], assessing driver drowsiness [9], and in the study of ergonomics [10]. Several studies have also been carried out on the development of robots controlled by eye movements [11–13].

The amplitude of the EOG signal varies with each person and environment; however, it is considered that it is in the range of 50–3500 µV. The amplitude of the signal obtained by placing two electrodes for registration is directly proportional to the angle of rotation of the eyes within the range of ±30°, as can be seen in Figure 1. Sensitivity is in the order of 20 µV per degree of movement [14,15]. The frequency ranges from continuous at about 50 Hz, although almost its entire spectrum, where most of the useful information resides, does not exceed 38 Hz. The fact that the EOG signal has a low bandwidth of interest is because the action potentials do not occur at extreme speed. Another interesting aspect to keep in mind is that muscle noise extends across the signal bandwidth almost steadily, which makes it very difficult to eliminate it in its entirety.

As exposed, the EOG signal is small in amplitude and consists of very low frequencies, so the presence of artifacts, interference, and noise in the biopotential recording is practically inevitable. They may occupy either some specific frequency band or the entire frequency band. Therefore, they are very difficult to remove without losing some signal information. The development of an Analog Front-End (AFE) for EOG recording is a challenging task and an active area of research; however, most work has not been developed or marketed as the BlueGain bioamplifier from Cambridge Research Systems [16] for a general purpose. The acquisition stage is usually oriented to a specific application, mainly for the control of human–computer interfaces, and oriented to individuals with amyotrophic lateral sclerosis [17,18]. Furthermore, most of these studies that can be found in the literature do not justify the design criteria followed in the proposed acquisition module. Step by step, this work exposes the design of an AFE with two differential analog input channels to record the horizontal and vertical movements of the eyes. The signals recorded by the two channels are amplified, filtered, digitized, and sent to the computer via Bluetooth. The alternatives and design criteria exposed in this work can be very useful for EOG-based equipment developers or researchers in this field.

The remainder of the article is structured as follows. Section 2 introduces two approaches to the acquisition of the EOG signal. Section 3 presents the external and internal sources of noise that affect the design of the EOG systems. In Section 4, the EOG AFE design is addressed for a general purpose. Finally, Section 5 provides conclusions.

## 2. Analog Front-End Approaches

Based on the resolution of the analog-to-digital converters (ADCs) used in the signal chain, there are two different approaches to the design of the AFE of an EOG system. The first one is to amplify the input signal significantly using low-noise amplifiers and a low-resolution ADC. The second approach would be to use a lower gain and a high-resolution ADC. In any case, the noise-free dynamic referred to as the input of the system is the same in both approaches.

### 2.1. AFE Based on Low-Resolution ADCs

Figure 2 shows a typical EOG AFE using a low-resolution ADC, typically 16-bit or less. It consists of two identical channels, the horizontal channel to acquire the biopotentials corresponding to the horizontal movements of the eyes and the vertical channel for the biopotentials relating to the vertical eye movements.

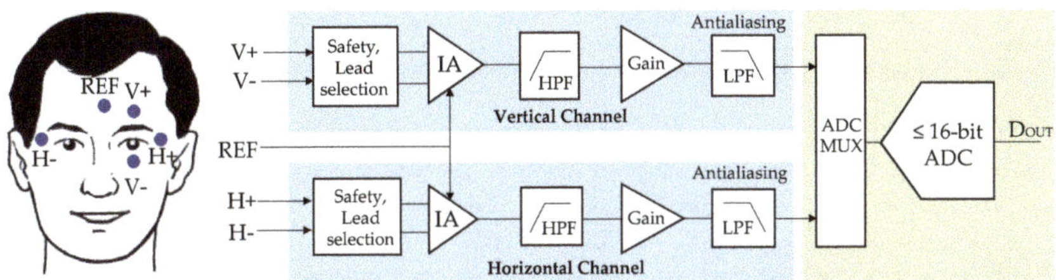


**Figure 2.** Typical EOG signal chain based on low-resolution analog-to-digital converters (ADCs). (REF: Reference; IA: Instrumentation amplifier; HPF: High-pass filter; LPF: Low-pass filter.)

In Figure 2, the first block from the left is intended for user protection, which could include high-value resistors or any other kind of isolation circuitry. The lead selection circuitry determines various electrode combinations that are possible to use. In Figure 2, five electrodes are shown; however, other configurations are possible, depending on the application and the required accuracy [19].

At low frequencies, electrodes behave as a high-impedance signal source, and therefore, they would need a high-impedance circuit for measuring the EOG signals [7]. This circuit is an instrumentation amplifier (IA) that has a very high common mode rejection ratio (CMRR) ($\geq$100 dB) and high input impedance (typically $10^{10}$ $\Omega$).

Before the EOG signal is passed to the ADC, it must be amplified so that the entire dynamic range of the ADC is used. For example, considering a 1 mV EOG input signal and an ADC input range from 0 to +2.5 V, the total gain would be 2500. This gain is distributed between the IA and an additional gain amplifier. Gain is added to the IA in such a way that the electrode DC offset does not saturate the input buffers of the IA. The actual value of this gain depends on the operating voltage of the IA. At this point, it is necessary to remove the DC component due to changes in the electrode–skin impedance with electrode motion. Typically, a second-order active-high-pass filter (HPF) with a corner frequency of 0.05 Hz is added to each signal chain.

After the DC component is removed, the signal is amplified again to achieve the total gain. For this issue, the operational amplifiers (op-amps) must be low-noise () to avoid a noisy system. Besides, for battery-powered systems, these amplifiers must be low-power. The drawback of the low power and low noise requirements is an increase in the cost of the precision op-amps.

The amplification stage is followed by a very sharp low-pass antialiasing filter to avoid out-of-band noise, produced by successive approximation register (SAR) ADCs. Typically, a fourth-order Chebyshev low-pass filter is used. This block is followed by a multiplexer (MUX) embedded into the ADC. As can be seen in this AFE approach, there is a significant amount of analog signal processing that occurs before the signal is digitized.

## 2.2. AFE Based on High-Resolution ADCs

Figure 3 shows the same EOG AFE with a high-resolution ADC implemented, typically a 24-bit sigma-delta (converter. This kind of ADC is characterized by a very high-resolution based on oversampling and noise-shaping principles.

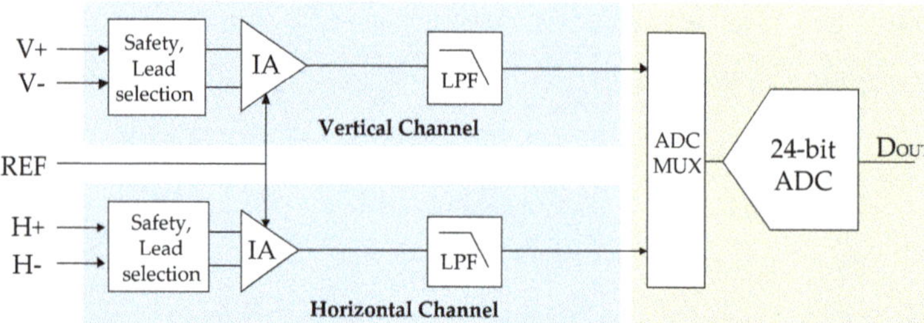


**Figure 3.** EOG AFE based on high-resolution ADC (Sequential Sampling).

In this approach, the first stage of the AFE is an instrumentation amplifier with a low gain in the range of 5 to 20. Because amplification is so low, the resolution of the ADC must be high, usually 24 bits, to digitize the EOG signal in the right way. Because the filtering is not done in the analog domain, a lot of processing is generally needed to be done in the digital domain. The noise referred to the input of the system depends on the output rate of the ADC and the gain of the IA but is usually within the commercial EOG requirements [16].

In this approach, the high-pass filter, DC blocking filter, amplification stage, and active low-pass filter are removed. The ADC has an integrated MUX and a programmable gain amplifier (PGA) that eliminates the need for signal condition circuitry. In addition, to offer the advantage of higher resolution, ADCs significantly relax the antialiasing requirements before the ADC, because the input is highly oversampled. Antialiasing filters can be replaced by a simple, single-pole 150 Hz, RC filter. On the other hand, the DC high-pass filter is removed because it can be implemented digitally. Using digital filtering would be possible to implement an adaptative DC removal filter to better reject the baseline wandering.

For most EOG applications, a sequential sampling may be acceptable. However, for certain applications such as ataxic disorder diagnosis [20] where the out-of-phase between channels is the most important feature, it is necessary to use a dedicated ADC for each channel.

In conclusion, a low-resolution ADC approach requires more analog signal processing and limited flexibility. However, a high-resolution ADC approach is associated with a higher power and a higher area. All the filtering and extra gain is done in the digital domain, which increases the processing requirements and power. To manage high-resolution ADCs, manufacturers offer evaluation boards and development software that can help with the design process significantly.

## 3. Noise Sources

EOG signals may be corrupted by various kinds of noise, such as electrode contact noise, power line interferences, motion artifacts, muscle contraction (EMG), baseline drift, intrinsic noise generated by electronic devices, electrosurgical noise, and other less significant noise sources. A correct understanding of each kind of noise is very important because it determines the resolution and the dynamic range of the EOG system. In this article, the main kinds of noise and artifacts are briefly commented on together with common solutions to minimize them using hardware techniques. A more complete description can be found in references [7,15,21].

### 3.1. Electrode Contact Noise

Electrodes are the first element in the measurement channel and play an important role in the design of an AFE circuit. Biopotential electrodes must covert the flow of ionic current into an electronic current [22]. This transduction must be done as faithfully as possible, and in addition, it must not disturb the EOG signal. The important parameters are, then, impedance and noise. The impedance should be as low as possible to reduce the charging effect of the subsequent amplification stage and minimize the effect of common-mode interference appearing at the input [23].

Currently, two types of electrodes are used to register the EOG signal: traditional surface electrodes and those known as dry electrodes. The main difference between dry and wet electrodes is that wet ones require pre-positioning preparation and dry ones do not. However, these have a much higher price than wet ones. For this reason, the first are today used the most in the clinical environment and the second, in research [24,25].

By placing a surface electrode in contact with the skin through an electrolyte, a distribution of charges occurs at the electrode–electrolyte interface, which results in the appearance of a potential called half-cell potential. If the electrode moves with respect to the electrolyte, there will be an alteration in the distribution of the charge, which will cause a transient variation in the half-cell potential [25,26]. In the same way, at the electrolyte–skin interface, there will also be a distribution of charges and, therefore, an equilibrium potential that will vary if there is movement between the skin and the electrolyte. Generally, the surface electrodes have a quite powerful adhesive that limits the displacements. Even so, the potential of the interfaces may vary due to the presence of the stratum corneum. This type of interference called a motion artifact produces a fluctuation in the signal at very low frequencies (<1 Hz), not susceptible to being filtered due to the large amount of information that the EOG signal possesses at these frequencies. In a differential amplification, the EOG signal will be superimposed on a direct voltage since these contact potentials will hardly be the same in both inputs. This limits the gain of the first amplifier stage since this DC voltage could saturate the amplifier. Figure 4 shows an example of the motion artifact and the real part of the electrode impedance change at 16 Hz [27,28].

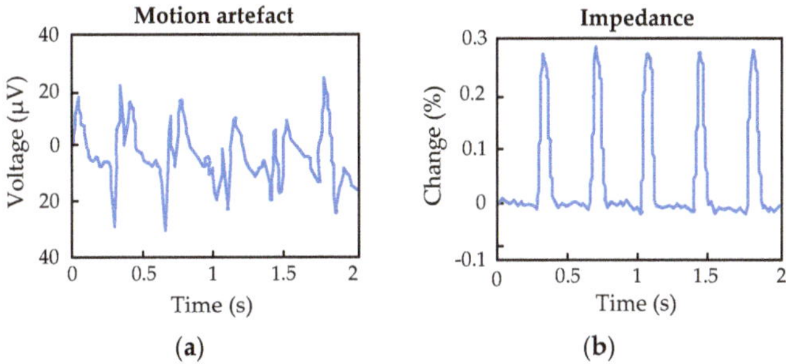


**Figure 4.** (a) Measurement of the motion artifact; (b) Real part of the electrode impedance at 16 Hz.

### 3.2. Noise from Other Biopotentials

In the human body, the different biopotential signals are not isolated. Throughout the body, electrical signals are constantly generated from different physiological systems. The most influential are muscle activity—the EMG—and the EEG signal since both are produced close to the eyes, and the bandwidth of the EOG signal acts within the bandwidth of those signs. Cardiac activity and retinal potential have less influence, the first being in a higher frequency band and the second having a smaller amplitude.

Disturbances caused by muscular actions, such as chewing, the opening or closing of the eyes, frowning, etc., considerably affectthe EOG signal recording [24–28]. The most important muscular action is blinking since it is involuntary, and although it does not modify the electrostatic potential of the eye, it can move the electrodes, creating electrode–skin interference. This causes the appearance of high-frequency artifacts. In addition, blinking (Figure 5) can be confused with a saccadic movement since it has the same frequency and amplitude and involves a reflex vertical movement of the eyeball [19]. Blinks can be detected during computerized processing, using mathematical tools such as wavelet transforms [29].

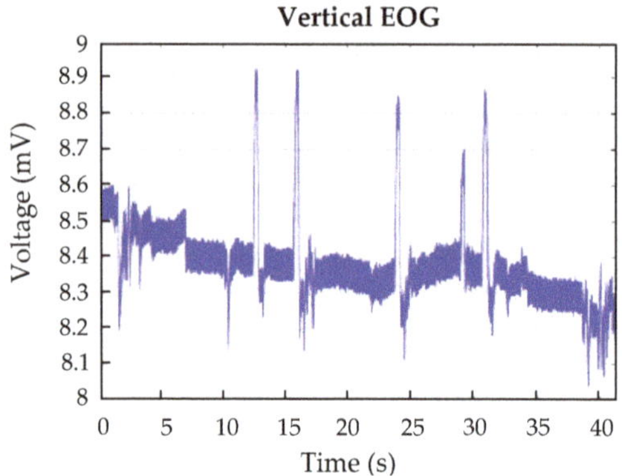


**Figure 5.** Example of blinking's effect on the vertical derivation.

*3.3. Power Line Interferences*

The main source of external interference is undoubtedly the electrical distribution network (50 Hz or 60 Hz) and its harmonics emitted by the conductors that run through the structure of a building. This leads to the appearance of electric and magnetic fields that interact with the measuring equipment and the user. Since these are low-frequency fields, they can be considered independent of each other since they will always be in the near field. Both will be minimized considering the design tips described in the following subsections.

### 3.3.1. Capacitive Interference

The capacitive coupling of the powerline is one of the main sources of interference present in the EOG registry. Figure 6 shows the human body connected to the measuring equipment by means of three electrodes including the capacitive interferences generated. The capacitive coupling with the body and the electrodes is the most influential aspect in the design of the biopotential amplifier circuit. This interference causes the appearance of displacement currents that flow to the ground through the electrode–skin interfaces and the patient's body. This also causes common-mode and differential voltages to appear at the amplifier input, resulting in a new source of interference. The capacities associated with this type of coupling range from 100 pF to 1 nF, depending on the user's position and the acquisition conditions. The capacities between the user and the powerline typically range from 0.2 pF to 20 pF. The capacitive coupling between the powerline and the AFE ($C_3$) does not cause interference since the interfering current that circulates through it goes to the ground through the housing of the equipment, therefore not entering into it. This interference can be very high in case of using very long cables located near the powerline or the equipment to which they are connected [15,30].

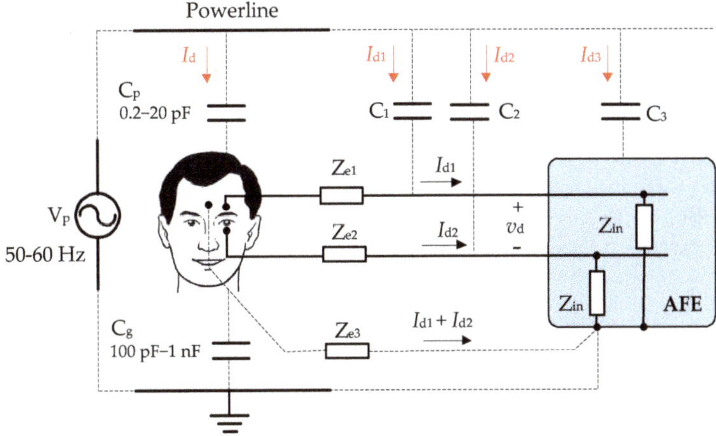

**Figure 6.** Capacitive coupling between the electrical network and the user.

### 3.3.2. Inductive Interference

The electrical current from the powerline produces a magnetic flux that crosses the loops formed in the measurement system, inducing noise voltages at 50 Hz or 60 Hz. The most important ones are induced in the loop formed by the patient, the drivers, and the measuring equipment [27].

If the loop is stationary and the flux density is sinusoidal, varying with time but constant in the area of the loop, the root mean square (RMS) value of the noise voltage is given by, where is the frequency in rad/s, is the effective value of the flux density, is the area of the loop, and is the angle that indicates the orientation of the measurement loop with respect to the magnetic field (see Figure 7a). The inductive voltage is always proportional to the frequency (in the capacitance it is only at low frequency) and now is independent of the input impedance of the measurement circuit (in the capacitance increase with this impedance). A very effective solution to reduce the induced noise voltage due to magnetic interferences is twisting the wires of the electrodes as shown in Figure 7b [31–33].

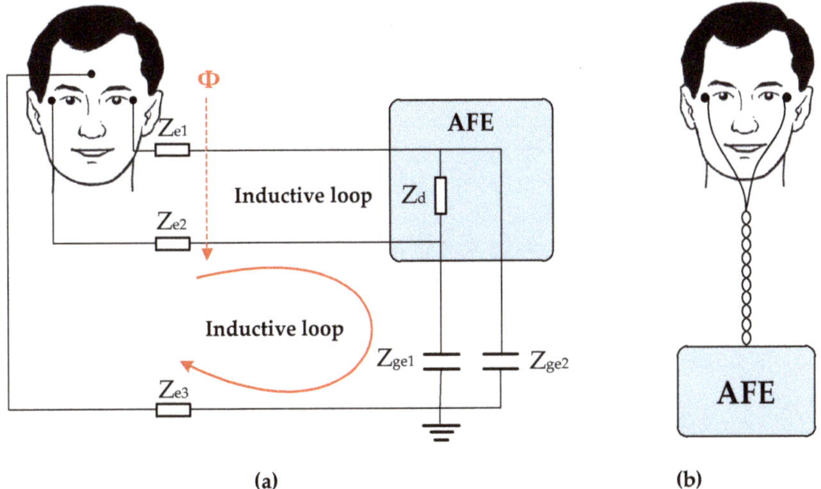

(a)  (b)

**Figure 7.** (a) Inductive loop present in the measurement system; (b) Reduction of magnetic interferences by twisting the wires of the electrodes.

### 3.4. Intrinsic Noise

Intrinsic noise is present in almost all electronic components, such as resistors and semiconductor devices [34]. The root-mean-square (rms) noise associated with the resistor can be estimated by where $R$ is the resistance, $T$ is the temperature in kelvins, $k$ is the Boltzmann's constant, and $B$ is the noise bandwidth in hertz. This expression illustrates the importance of using low resistance components when possible in low-noise circuits. For example, the noise in a 100 k$\Omega$ resistor at 25 °C (298 K) over the range of 0.1 Hz to 100 Hz is 0.4 µV. Large resistors are used as the input resistor of an op-amp gain circuit; their thermal noise will be amplified by the gain in the circuit. Thermal noise in resistors is often a problem in portable equipment, where resistors have been scaled up to get power consumption down.

On the other hand, noise for op-amps is usually specified with a graph of equivalent input noise versus frequency. As shown in Figure 8, these graphs usually show two distinct regions: lower frequencies where pink noise is the dominant effect, and higher frequencies where 1/f noise (white noise) is the dominant effect. The point in the frequency spectrum where 1/f noise and white noise are equal is referred to as the noise corner frequency, $f_{ce}$. In this way, the noise at the input of an op-amp can be estimated by the expressions: where is the voltage (current) white noise. An identical expression exists for the rms current noise.

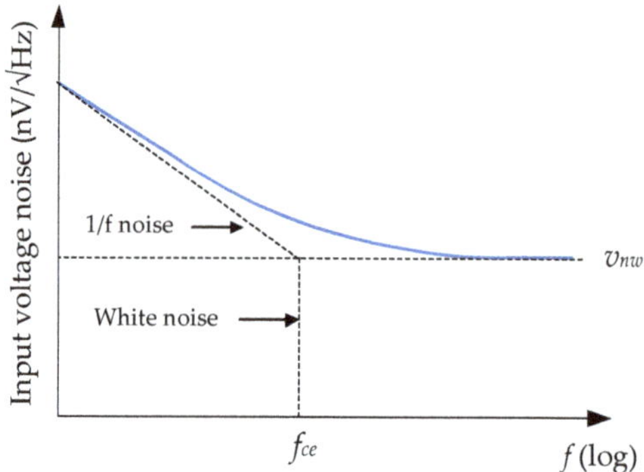


**Figure 8.** Input voltage noise spectral density vs. frequency [35].

In summary, in low-noise design, the following must be considered:

- Select op-amps with low-noise floors as well as low corner frequencies.
- Keep the external resistances sufficiently small to make the current noise and thermal noise negligible compared to the voltage noise.
- Limit the noise-gain bandwidth to the strict minimum required.

### 3.5. Baseline Drift

The information relative to the ocular position within the orbit of the eye is given by the continuous component due to the linearity of the behavior of the dipole of the eye in the ±30° range. However, the EOG signal has a strong variability or displacement of its continuous component on the isoelectric baseline. It is known as offset or drift [36] and is produced by different factors. It also has a huge variability from one person to another [7]. The baseline drift is slower than the eye movement, and it manifests itself as a rising or declining slope in the EOG waveform, as can be seen in Figure 5.

The main reason for this effect is given by the ambient brightness and the variations in the ocular dipole potential produced by the polarization differences versus different light intensities of the two photoreceptors of the retina (cones and rods). The lengthy attachment of the adhesive electrodes to the skin also causes baseline drift on the recorded EOG signal. This causes errors in the determination of the ocular position [7]. This drift can lead to the saturation of the amplifiers due to its high gain. The level of drift can rise up to a range of 0.2 V in an interval of a minute [37].

Most interference is removed in the hardware device; however, wideband noises and the baseline drift are not easily suppressed in the hardware stage. It is more effective to remove them using mathematical tools. The EOG signal is non-stationary (its spectrum varies over time) in such a way that many of its temporal aspects cannot be adequately analyzed with the Fourier transform, or with the Fourier transform with window. For these types of signals, the Wavelet transform can concentrate better on transient and high-frequency phenomena [38], which provide a better understanding of the EOG signal [29].

## 4. Results

Having presented the different approaches in Section 2 and the main sources of noise in Section 3, we are ready to design a prototype of an AFE based on the low-resolution ADC approach.

The EOG signals are picked up by an AFE system consisting of horizontal and vertical channels as shown in Figure 9. Both AFE channels would have the same design, but, for simplicity, only one AFE is represented. The acquisition system employs Ag/AgCl surface electrodes for signal pickup, which includes electrolyte gel to reduce contact impedance.

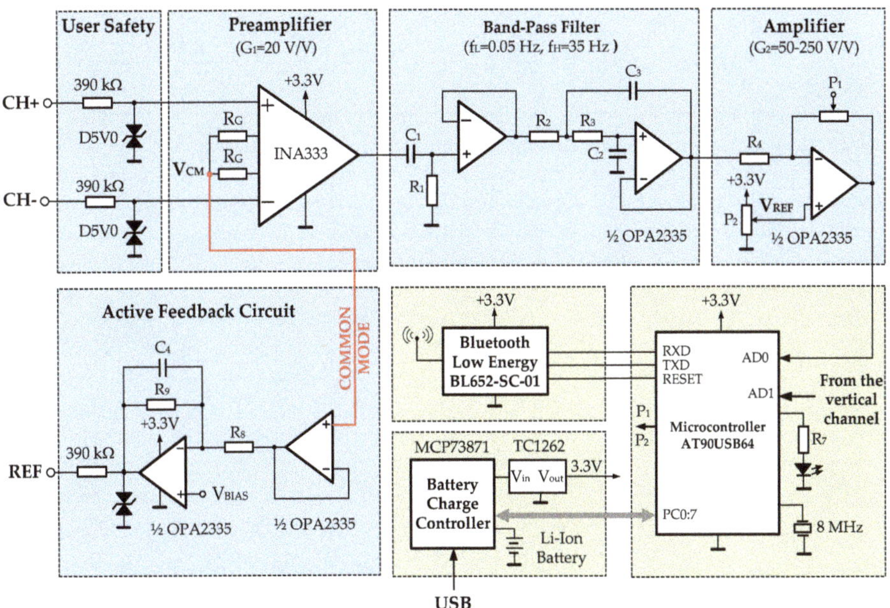


**Figure 9.** The electrical diagram of each channel of the developed system.

### 4.1. User Safety

The safety of the user and the device is the main issue. The designer must consider all scenarios and meet regulations. The most important standard to consider is IEC-60601 [39,40] which limits the current through the electrodes to less than 10 μA rms. To ensure this limit, a 390 kΩ resistance is placed in the signal path. It is widely used as a low cost approach.

To protect the user against transient overvoltages, a transient suppressor diode (D5VO) between each electrode and ground is placed. These diodes operate suppressing all overvoltages above the breakdown voltage and shunt excess current. In medical applications, an isolated amplifier is usually placed between the IA and the filter stage. For example, ISO124 is a low-cost precision isolation amplifier for which no external components are required for operation.

*4.2. Amplification*

The AFE can readily be designed using the instrumentation amplifier (in-amp) as the main component. For example, considering a 1 mV EOG input signal and an ADC input range from 0 to +2.5 V, the total gain would be 2500. This gain is distributed between the in-amp (INA333) and an additional gain amplifier. The INA333 is a micro-power, zero-drift, and rail-to-rail in-amp. It is characterized by a very low offset voltage (typ. 10 µV) and high common-mode rejection (typically 115 dB). It operates with power supplies as low as 1.8 V (±0.9 V), and the quiescent current is only 50 µA—ideal for battery-operated systems, like this design.

Gain is added to the INA in such a way that the DC electrode offset does not saturate the INA. The INA333 is configured to a gain of $G = 1 + 100\ \text{k}\Omega/R_G = 10\ v/v$ with an external 0.1% $R_G = 11.1\ \text{k}\Omega$ resistor. The last amplification stage simply amplifies the filtered signal to the desired output voltage range for the AFE (0–5 V). Its gain is set-up to by the digital potentiometer $P_1$ (Microchip MCP4161), controlled via the microcontroller serial peripheral interface (SPI). Another digital potentiometer, $P_2$, is used to adjust the reference voltage of each channel.

*4.3. Filtering*

In this field, it is common to use a high-pass filter to eliminate the continuous component and, therefore, its variability. It should be pointed out here that the variables measured in the human body (any biopotential) are rarely deterministic. Its magnitude varies with time, even when all possible variables are controlled. This means that the variability of the EOG reading depends on many factors that are difficult to determine: interferences caused by other biopotentials such as EEG and EMG, and those due to the positioning of the electrodes, skin–electrode contacts, head, and facial movements, lighting conditions, blinking, etc. In this case, to avoid this problem a high-pass filter with a cut-off frequency at 0.05 Hz and a relatively long time constant is used.

Different types, specifications, and responses of analog filters are used to limit the bandwidth of each stage. Before the biopotentials are amplified significantly, DC potentials from the electrode–skin interface must be filtered, to avoid the amplifier being saturated. In Figure 9, a high-pass filter formed by $C_1$ and resistor $R_1$ is set to 0.05 Hz. This filter could be distorting the biopotential being measured due to the artifacts of movements. In that case, mathematical tools such as the wavelet transform are used to identify and remove artifacts so they do not affect the behavior of the system [29]. The biggest problem in biopotential measurement is the interference from the powerline. Above all, RF interferences and muscle signal interference must be attenuated. For these aims, a second-order, Butterworth, 35 Hz low-pass filtering is used. Sometimes, it may be desirable to include a 50 Hz or 60 Hz notch filter to remove the power line interference, although it is possible the biopotential signal could be distorted.

*4.4. Active Feedback Circuit*

The common-mode interference is principally rejected by an instrumentation amplifier with a CMRR of a minimum of 100 dB. Further improvement is possible by using a circuit to actively cancel the interference, named in Figure 4 as an active feedback circuit (AFC). This circuit employs the same idea used in ECG systems. The common-mode signal is sensed from the first stage of the instrumentation amplifier, amplified, inverted, and fed back into the reference electrode. At this stage, the common-mode signal is reduced by the term $(1 + 2\ R_9/R_G)$ [7,15]. The AFC circuit along with a high CMRR of the amplifier permits very high-quality biopotential measurements.

### 4.5. Additional Features

The last gain stage is followed by a multiplexer block that feeds into a 10-bit ADC, both embedded in the microcontroller device (AT90USB1287). It is a low-power 8-bit microcontroller based on RISC architecture, 128 Kbytes of ISP Flash, and a USB controller.

The frequency range of the EOG signal is typically between 0 and 50 Hz. To fulfill the Nyquist criterion, the sampling frequency must be at least 100 samples per second. The ADC output is carried to the Bluetooth Low Energy device (BL652-SC-01 from Laird Connectivity), through the microcontroller USART (Universal Synchronous and Asynchronous serial Receiver and Transmitter). The sample rate at which each channel is scanned by the ADC is 9600 bps.

A 3.3 V power supply is obtained from 5 V associated with the USB connector by means of a low dropout regulator (Microchip TC1262). The circuitry is power supplied by a 3.7 V/2000 mAh Li-ion battery, which is recharged via a Microchip MCP73871 controller. The power consumption was estimated by dividing the battery capacity (mAh) by the average current consumed by the device. It was estimated that the average battery life was about ninety hours.

### 4.6. PCB Layout

A high-performance EOG system requires a careful PCB layout. The first aspect to consider is the separation of the analog and digital sections of the PCB. This keeps the noisy digital from the low-level analog circuits. Because at high frequencies, the return current takes the path of least inductance, a low-inductance ground is necessary. The ground solid plane provides a low inductance return for signal current. Discontinuities on the ground plane (slots or splits) produce large current loops that increase the inductance of the ground plane.

Related to power supply distribution, the decoupling capacitor must be used close to the $V_{DD}$-GND pins. They provide a source of charge when the IC switches and offer low AC impedance between the power and ground rails. For these functions, multilayer ceramic chip capacitors (MLCC) are chosen due to their best relation of capacity to size.

Critical signal traces (clock, buses, and control signals) should be routed first and away from the edge of the PCB. These signals contain high amplitude harmonics. The clock traces should be kept as short as possible, and optimum placement should be provided by routing them first. To reduce cross-coupling effects when the analog and digital signals are mixed, the lines should be made to cross each other at 90-degree angles [15,21]. To reduce the inductive (capacitive) couplings, the long parallel traces on the same layer (on adjacent layers) should be minimized.

The EOG system is implemented on a 50 mm × 100 mm PCB. This prototype is a cost-effective solution (around USD 80, including the PCB and electrodes). Figure 10a shows a photo of the device with the leads. Figure 10b shows the waveforms obtained at the output of the vertical channel and horizontal channel, and Figure 10c shows the saccadic movements corresponding to a 0°–10°–20°–30°–40° gaze sequence. Figure 10c was obtained by limiting the bandwidth of the oscilloscope to filter the noise. The device has been tested by twenty-five people, obtaining a similar signal quality in all cases. As can be seen, the signals still require more computational processing to clean the signal.

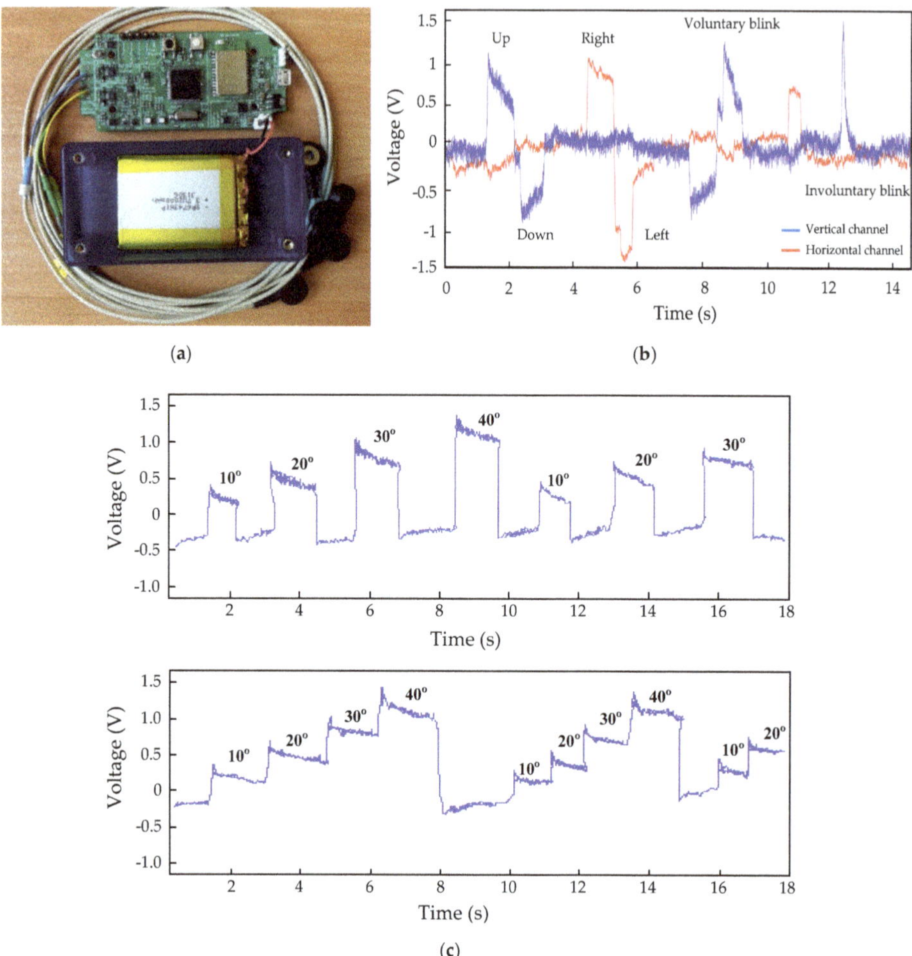



**Figure 10.** (**a**) A photo of the EOG system; (**b**) Characteristic EOG potential changes due to eye movement; (**c**) Saccadic movements corresponding to a 0°–10°–20°–30°–40° gaze sequence.

Finally, Table 1 highlights the main differences between a reference EOG biosignal amplifier (BlueGain from Cambridge Research Systems) [16] and the proposed EOG device. As can be seen in Table 1, the figures for the resolution and sample rate of the commercial device are higher than those of the proposed EOG device. However, the resolution is limited by the noise; therefore, a resolution of nanovolts might be difficult to achieve. The sample rate of 1 kHz is enough for this type of signal, which fulfills the Nyquist criterium. On the other hand, the battery life and the cost are two important features that have been optimized by using a Bluetooth low energy module and low-cost components.

**Table 1.** Comparison of characteristics between the BlueGain Amplifier and the proposed device.

| | BlueGain EOG-Amplifier | Proposed EOG Device |
|---|---|---|
| | Analog Characteristics | |
| Input Voltage Range | 0 to 80 mV | 0 to 50 mV |
| Frequency Response | DC to 150 Hz | 0.05 Hz to 35 Hz |
| CMRR | 110 dB | 115 dB |
| Battery life | Typically, 75 h with lithium AA Cells (2 Required) | 90 h with a 3.7 V/2000 mAh Li-ion battery |
| Digitization | 16 bits | 10 bits |
| Sample Rate | 10 kHz max. | 1 kHz max. |
| Resolution | 122 nV | 1 mV |
| | Digital Characteristics | |
| Bluetooth | Yes. Not specified | Low Energy |
| Service | Virtual Serial Port, SLIP | Virtual Serial Port |
| Infra-red marker channel | 0–5 V | No |
| | Physical Characteristics | |
| Weight | 157 g | 100 g |
| Patient Connection | 1.5 mm touch proof connectors | Connectors embedded in the enclosure |

## 5. Conclusions

This paper is focused on the characteristics of electrocardiographic signals and the different front-end approaches for EOG signal acquisition. The tradeoffs between different approaches and the effects on overall system design are discussed. EOG signals may be corrupted by various kinds of noise, and it must be filtered, which can distort the signal, requiring substantial computational postprocessing. All these issues are important design considerations for applications in real-time monitoring and were discussed in this work. To validate the study, a prototype of the EOG system was implemented in a 10 × 5 cm PCB. The PCB layout is a crucial issue for the functionality and electromagnetic compatibility performance of an EOG system.

For future work, signal processing provides a great deal of flexibility. This could be beneficial to using a high-resolution ADC, moving the signal processing to the digital domain. Therefore, in the future, it would be interesting to redesign the device using a high-resolution ADC. Once the EOG signal has been processed properly, this portable and low-cost device could be used for controlling devices such as virtual keyboards, powered wheelchairs, robots, and medical applications, etc. An important field of application is the development of assistive technology geared towards various disabled people through the development of human–computer interfaces. These kinds of people retain control of their eye movements, which can be translated into commands.

**Author Contributions:** Conceptualization, A.L. and F.F.; methodology, F.F.; software, J.R.V.; validation, A.L. and F.F.; formal analysis, O.P.; investigation, A.L.; resources, A.L. and F.F.; data curation, O.P.; writing—original draft preparation, A.L. and F.F.; writing—review and editing, F.F.; visualization, J.R.V.; supervision, O.P.; project administration, A.L.; funding acquisition, A.L. and F.F. All authors have read and agreed to the published version of the manuscript.

**Funding:** This research has been funded by the Spanish Ministry of Science and Innovation under project MINECO-TIN2017-84804-R and by the project grant FC-GRUPIN-IDI/2018/000226 from the Asturias Regional Government.

**Conflicts of Interest:** The authors declare no conflict of interest.

## References


1. Malmivuo, J.; Plonsey, R. *Bioelectromagnetism, Principles and Applications of Bioelectric and Biomagnetic Fields*; Oxford University Press: New York, NY, USA, 1995; Chapter 28.
2. Knapp, R.B.; Lusted, H. Biological signal processing in virtual reality applications. In Proceedings of the 4th International Conference of the Virtual Reality and Persons with Disabilities, San Francisco, CA, USA, 22–25 June 1993; pp. 134–137.
3. Liang, S.F.; Kuo, C.E.; Lee, Y.C.; Lin, W.C.; Liu, Y.C.; Chen, P.Y.; Cherng, F.Y.; Shaw, F.Z. Development of an EOG-based automatic sleep-monitoring eye mask. *IEEE Trans. Instrum. Meas.* **2015**, *64*, 2977–2985. [CrossRef]
4. Pal, M.; Banerjee, A.; Datta, S.; Konar, A.; Tibarewala, D.N.; Janarthanan, R. Electrooculography based blink detection to prevent computer vision syndrome. In Proceedings of the International Conference on Electronics, Computing and Communication Technologies (CONECCT'14), Bangalore, India, 6–7 January 2014; pp. 1–6.
5. García-Bermúdez, R.; Ruiz, F.R.; Peñalver, J.G.; Cansino, O.V.; Pérez, L.V.; Torres, C.; Becerra-García, R. Evaluation of Electro-oculography data for Ataxia SCA-2 classification. In Proceedings of the 10th International Conference on Intelligent Systems Design and Applications (ISDA'10), Cairo, Egypt, 29 November–1 December 2010; pp. 237–241.
6. Becerra, R. Saccadic points classification using multilayer perceptron and random forest classifiers in EOG recordings of patients with ataxia SCA2. In Proceedings of the 12th International Work-Conference on Artificial Neural Networks (IWANN), Puerto de la Cruz, Tenerife, Spain, 12–14 June 2013; pp. 115–123.
7. U.S. National Library of Medicine. The Medline Plus Merrian-Webster Medical Dictionary. Available online: http://medlineplus.gov/ (accessed on 11 February 2020).
8. uvinage, M.; Cubeta, J.; Castermans, T.; Petieau, M.; Hoellinger, T.; Cheron, G.; Dutoit, T. A quantitative comparison of the most sophisticated EOG-based eye movement recognition techniques. In Proceedings of the IEEE Symposium on Computational Intelligence, Cognitive Algorithms, Mind and Brain (CCMB'13), Yew-Soon Ong, Singapore, 16–19 April 2013; pp. 44–52.
9. Ebrahim, P.; Stolzmann, W.; Yang, B. Eye movement detection for assessing driver drowsiness by electrooculography. In Proceedings of the IEEE International Conference on Systems, Man, and Cybernetics, San Antonio, TX, USA, 13–16 October 2013; pp. 4142–4148.
10. Schleicher, R.; Galley, N.; Briest, S.; Galley, L. Blinks and saccades as indicators of fatigue in sleepiness warnings: Looking tired? *Ergonomics* **2008**, *51*, 982–1010. [CrossRef] [PubMed]
11. Nam, Y.; Koo, B.; Cichocki, A.; Choi, S. GOM-face: GKP, EOG, and EMG-based multimodal interface with application to humanoid robot control. *IEEE Trans. Biomed. Eng.* **2014**, *61*, 453–462. [CrossRef] [PubMed]
12. Ma, J.; Zhang, Y.; Cichocki, A.; Matsuno, F. A novel EOG/EEG hybrid human–machine interface adopting eye movements and ERPs: Application to robot control. *IEEE Trans. Biomed. Eng.* **2015**, *62*, 876–889. [CrossRef] [PubMed]
13. Úbeda, A.; Iáñez, E.; Azorín, J.M. An integrated electrooculography and desktop input bimodal interface to support robotic arm control. *IEEE Trans. Human-Mach. Syst.* **2013**, *43*, 338–342. [CrossRef]
14. Cohen, A. *Biomedical Signal Processing*; CRC Press: Boca Raton, FL, USA, 1986; Volume 1.
15. Webster, J.G. *Medical Instrumentation: Application and Design*; John Wiley & Sons Inc.: Hoboken, NJ, USA, 2010.
16. BlueGain Cambridge Research Systems. BlueGain EOG Biosignal Amplifier. Available online: http://www.crsltd.com/tools-for-vision-science/eye-tracking/bluegain-eog-biosignal-amplifier/ (accessed on 11 March 2020).
17. Larson, A.; Herrera, J.; George, K.; Matthews, A. Electrooculography based electronic communication device for individuals with ALS. In Proceedings of the IEEE Sensors Applications Symposium (SAS), Glassboro, NJ, USA, 13–15 March 2017; pp. 4142–4148.
18. Heo, J.; Yoon, H.; Suk, K. A novel wearable forehead EOG measurement system for human computer interfaces. *Sensors* **2017**, *17*, 1485. [CrossRef] [PubMed]
19. López, A.; Ferrero, F.J.; Valledor, M.; Campo, J.C.; Postolache, O. A study on electrode placement in EOG systems for medical applications. In Proceedings of the IEEE International Symposium on Medical Measurements and Applications (MeMeA), Benevento, Italy, 12–14 May 2016; pp. 29–33.

20. López, A.; Ferrero, F.J.; Postolache, O. An Affordable Method for Evaluation of Ataxic Disorders Based on Electrooculography. *Sensors* **2019**, *19*, 3756. [CrossRef] [PubMed]
21. Thakor, N.V. Biopotentials and electrophysiology measurement. In *The Measurement, Instrumentation and Sensors, Handbook*; Webster, J.G., Ed.; CRC Press: Boca Raton, FL, USA, 1999; Volume 74.
22. Carim, H. Bioelectrodes. In *Encyclopedia of Medical Devices and Instrumentation*; Webster, J.G., Ed.; Wiley: New York, NY, USA, 1988; pp. 195–226.
23. McAdams, E.T. Bioelectrodes. In *Encyclopedia of Medical Devices and Instrumentation*, 2nd ed.; John, G., Ed.; Wiley InterScience: New York, NY, USA, 2006; pp. 120–166. ISBN 0-471-26358-3.
24. Srinivas, M.G.; Pandian, P.S. Dry electrodes for bio-potential measurement in wearable systems. In Proceedings of the 2nd IEEE International Conference on Recent Trends in Electronics, Information & Communication Technology (RTEICT), Bangalore, India, 19–20 May 2017; pp. 270–276.
25. Yokus, M.A.; Jur, J.S. Fabric-based wearable dry electrodes for body surface biopotential recording. *IEEE Trans. Biomed. Eng.* **2016**, *63*, 423–430. [CrossRef] [PubMed]
26. Fernández, M.; Pallás-Areny, R. Electrode Contact Noise in Surface Biopotential Measurements. In Proceedings of the 14th Annual International Conference of the IEEE Engineering in Medicine and Biology Society (EMBS), Paris, Francia, 29 October–1 November 1992; pp. 123–124.
27. Ko, B.H.; Lee, T.; Choi, C.; Kim, Y.H.; Park, G.; Kang, K.; Bae, S.K.; Shin, K. Motion artifact reduction in electrocardiogram using adaptive filtering based on half-cell potential monitoring. In Proceedings of the Annual International Conference of the IEEE Engineering in Medicine and Biology Society (EMBS), San Diego, CA, USA, 28 August–1 September 2012; pp. 1–7.
28. Simakov, A.B.; Webster, J. Motion artifact from electrodes and cables. *Iran. J. Electron. Comp. Eng.* **2010**, *9*, 139–143.
29. Bulling, A.; Ward, J.A.; Gellersen, H.; Tröster, G. Eye movement analysis for activity recognition using electrooculography. *IEEE Trans. Pattern Anal. Mach. Intell.* **2011**, *33*, 741–753. [CrossRef]
30. Pallás-Areny, R. Interference-rejection characteristics of biopotencial amplifiers: A comparative analysis. *IEEE Trans. Biomed. Eng.* **1988**, *35*, 953–959. [CrossRef] [PubMed]
31. Rosell, J.; Colominas, P.; Riu, R.; Pallás-Areny, R.; Webster, J.G. Skin impedance from 1 Hz to 1 MHz. *IEEE Trans. Biomed. Eng.* **1988**, *35*, 649–651. [CrossRef] [PubMed]
32. Frank, U.A.; Londer, R.T. The hospital electromagnetic environment. *J. Assoc. Adv. Med. Inst.* **1971**, *5*, 246–254.
33. Hubta, J.C.; Webster, J.G. 60 Hz interference in electrocardiography. *IEEE Trans. Biomed. Eng.* **1973**, *20*, 91–101.
34. Motchenbacher, C.D.; Connelly, J.A. *Low-Noise Electronic System Design*; John Wiley & Sons, Inc.: Hoboken, NJ, USA, 1993.
35. Baker, B.B. Matching the noise performance of the operational amplifier to the ADC. *Analog. Appl. J. Texas Instrum.* **2006**, *1*, 5–9.
36. Manabe, H.; Fukumoto, M.; Yagi, T. Automatic drift calibration for EOG-based gaze input interface. In Proceedings of the 35th Annual International Conference of the IEEE Engineering in Medicine and Biology Society (EMBS), Osaka, Japan, 3–7 July 2013; pp. 53–56.
37. Estrany, B.; Fuster, P.; García, A.; Luo, Y. Human computer interface by EOG tracking. In Proceedings of the 1st International Conference on Pervasive Technologies Related to Assistive Environments (PETRA), Athens, Greece, 16–19 July 2008; pp. 1–9.
38. Drongelen, W. *Signal Processing for Neuroscientists*; Academic Press/Elsevier: Amsterdam, The Netherlands, 2018; Chapter 15.
39. Brown, M.; Marmor, M.; Zrenner, E.; Brigell, M.; Bach, M. ISCEV Standard for Clinical. Electro-Oculography (EOG). *Doc. Ophthalmol.* **2006**, *113*, 205–212. [CrossRef] [PubMed]
40. International Organization for Standardization (ISO). Occupational Health and Safety Management Systems. Requirements with Guidance for Use. Available online: http://www.iso.org/iso/catalogue_detail.htm?csnumber=65529 (accessed on 15 March 2020).

 *electronics*

*Article*

# MEDUSA: A Low-Cost, 16-Channel Neuromodulation Platform with Arbitrary Waveform Generation

Fnu Tala and Benjamin C. Johnson *

Department of Electrical and Computer Engineering, College of Engineering, Boise State University, 1910 W University Dr, Boise, ID 83706, USA; fnutala@u.boisestate.edu
* Correspondence: bcjohnson@boisestate.edu; Tel.: +1-208-426-3842

Received: 17 March 2020; Accepted: 7 May 2020; Published: 15 May 2020

**Abstract:** Neural stimulation systems are used to modulate electrically excitable tissue to interrogate neural circuit function or provide therapeutic benefit. Conventional stimulation systems are expensive and limited in functionality to standard stimulation waveforms, and they are bad for high frequency stimulation. We present MEDUSA, a system that enables new research applications that can leverage multi-channel, arbitrary stimulation waveforms. MEDUSA is low cost and uses commercially available components for widespread adoption. MEDUSA is comprised of a PC interface, an FPGA for precise timing control, and eight bipolar current sources that can each address up to 16 electrodes. The current sources have a resolution of 15.3 nA and can provide $\pm 5$ mA with $\pm 5$ V compliance. We demonstrate charge-balancing techniques in vitro using a custom microelectrode. An in vivo strength-duration curve for earthworm nerve activation is also constructed using MEDUSA. MEDUSA is a multi-functional neuroscience research tool for electroplating microelectrodes, performing electrical impedance spectroscopy, and examining novel neural stimulation protocols.

**Keywords:** neuromodulation; multi-channel; stimulation protocol; FPGA

## 1. Introduction

Electrical stimulation is a broadly applied technique in neuroscience research and bioelectronic clinical therapies. Direct electrical stimulation of peripheral nerves has been shown to modulate physiological functions such as blood pressure [1] and rheumatoid arthritis [2]. In the central nervous system, electrical stimulation has been used to treat chronic pain via spinal cord stimulation [3] and alleviate the symptoms of Parkinson's disease with deep-brain stimulation [4]. Most neuromodulation is performed with charge-balanced, rectangular biphasic pulses [5] or monophasic pulses with passive recharge in order to conserve energy in an implanted device [6]. However, conventional electrical stimulation through a microelectrode has the highest current density at the electrode-tissue interface, meaning neural activation is maximized at the interface. Therefore, recent work has focused on utilizing novel stimulation waveforms to focalize electrical stimulus deeper in tissue away from the electrode interface. One proposed method is to use temporally interfering sinusoids with a small frequency offset (e.g., 10 Hz) and relatively high carrier frequency ($\geq$1 kHz) so as to not excite tissue near the electrode interface [7]. Prior work has shown cell firing entrained to the low offset frequency deep into tissue where the sinusoids interfere. Relatedly, another method is to use multiple intersectional short pulses from several electrode pairs to focus the stimulus [8]. This technique relies on the stimulus-duration relationship of excitable cells and many time-offset electrode pairs, where the targeted cell will integrate several subthreshold stimulation pulses. Novel stimulation paradigms also look to maximum stimulation efficiency to minimize energy dissipated by implantable devices. Overall, stimulation efficiency is the product of the efficiency of the electronics and how effective the stimulation waveform is at inducing the desired physiological response. For example, exponential waveforms are less energy efficient than rectangular waveforms in activating neurons and axons, but systems with exponential waveforms may be more efficient overall due to the way they can be generated with electronics [9].

While many ASIC implementations boast a powerful stimulation feature set [10–13], they are not widely available for low-cost neuroscience research. Since kilohertz stimulation frequencies have only recently garnered more interest, conventional neuromodulation equipment is not suitable for high frequency stimulation [14]. Furthermore, expensive benchtop equipment makes multi-channel or highly parallelized experiments impractical. As such, there is growing interest and research need in creating low-cost neuromodulation hardware [13,15]. In this work, we present MEDUSA, a low-cost platform capable of generating arbitrary, current-controlled stimulus waveforms. The platform can create high-resolution (<16 nA) currents of ±5 mA with ±5 V compliance. Additionally, the sinusoidal stimulus can be used for electrical impedance spectroscopy [16,17] or kilohertz frequency nerve block. This paper describes the architecture and characterization of the implemented system.

## 2. Materials and Methods

### 2.1. MEDUSA System Architecture

The implemented system architecture and system components are shown in Figure 1. MEDUSA was comprised of an FPGA/USB interface (XEM6010, Opal Kelly, Portland, OR, USA), an 8-channel, 16-bit DAC (DAC81408, Texas Instruments, Dallas, TX, USA), 8 voltage-controlled bipolar current sources that can address 16 channels through a 16:1 multiplexer, and a shorting switch for each channel. Users could configure stimulation waveform parameters through a Python interface on the PC. The settings were transferred via USB, and precise stimulation timing control was facilitated by the FPGA. The FPGA communicated with the DAC via SPI (serial peripheral interface) operating at 24 MHz and an asynchronous trigger signal (DAC_TRIG) that updated the DAC outputs. The grounding switches, controlled by the FPGA, could be used as a current return or used to clear residual charge from the electrode interface after stimulation. The multiplexers and grounding switches were selected for low leakage currents (<10 nA) and suitable low on-resistance (250–300 Ω).

### 2.2. Bipolar Current Source

The voltage-controlled bipolar current sources in Figure 2 used an amplifier with auxiliary differential inputs (LT6552, Linear Technology, Milpitas, CA, USA), and either current source was able to sink or source current. The auxiliary input was used in a feedback loop to force the equivalent voltage of $V_{DAC}$ across the output resistor $R_{GAIN}$. Therefore, in order to balance the current sinking and sourcing, the input voltage $V_{DAC1}$ and $V_{DAC2}$ must be differential. The resulting output current for each channel is given as:

$$I_{DAC} = V_{DAC1}/R_{GAIN1} = -V_{DAC2}/R_{GAIN2} \tag{1}$$

where $R_{GAIN<X>}$ was switchable for each current source to 100 Ω, 1 kΩ, or 10 kΩ and was chosen on resolution and dynamic range requirements.

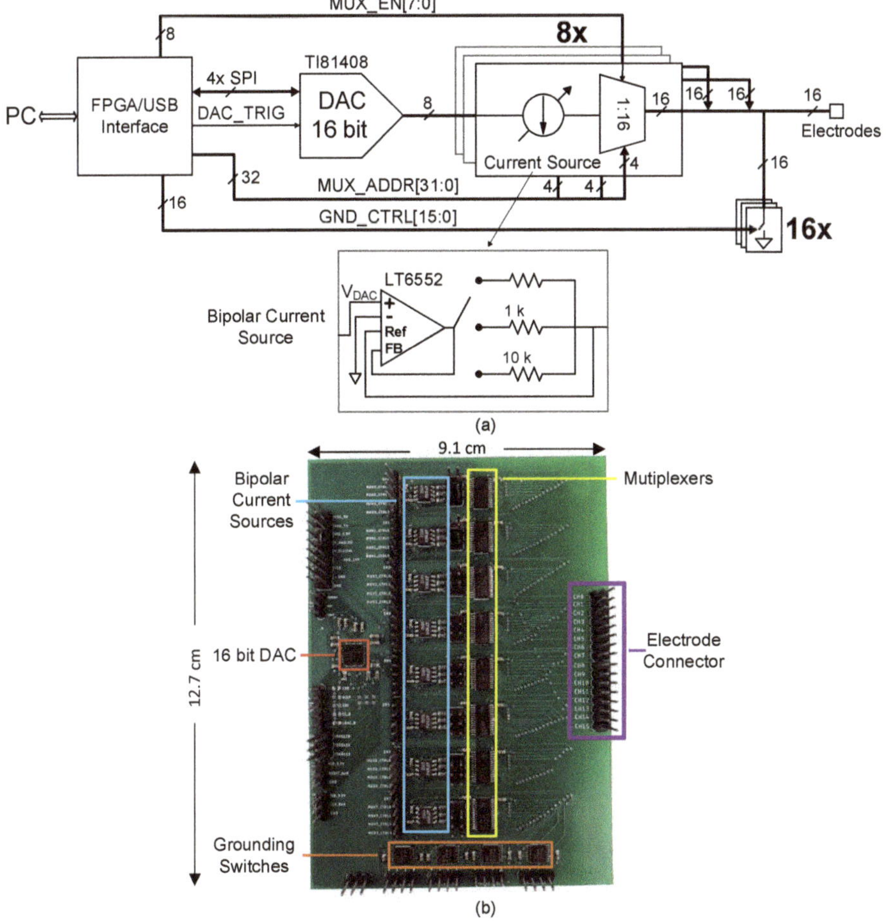


Figure 1. (a) System architecture and (b) implemented system.

Since the amplifier output could swing rail-to-rail ($\pm 5$ V) and source or sink 70 mA, the maximum current output was limited by the input common-mode of the auxiliary amplifier input (VSS to VDD − 1.4 V). The amplifier could use an asymmetric supply (VSS = −5 V, VDD = 7 V) to extend the voltage compliance range at the expense of current source linearity. The total voltage supply required for an anodic stimulation pulse could be calculated to be:

$$VDD_{MIN} \approx I_{DAC} \cdot (R_{GAIN} + R_S + R_{MUX} + T_P/C_I) + 1.4\ \text{V} \tag{2}$$

where $R_S$ is the spread resistance of the electrode, $R_{MUX}$ is the on-resistance of the multiplexer ($\approx 100\ \Omega$), $T_P$ is the stimulation pulse width, and $C_I$ is the interface capacitance of the electrode. The additional 1.4 V of headroom was required by the amplifier for anodic stimulation due to input common mode limitations. Note that this bipolar current source could also be used for unipolar current stimulation, since this system had active grounding switches, so the second electrode could be short to ground and made it unipolar.

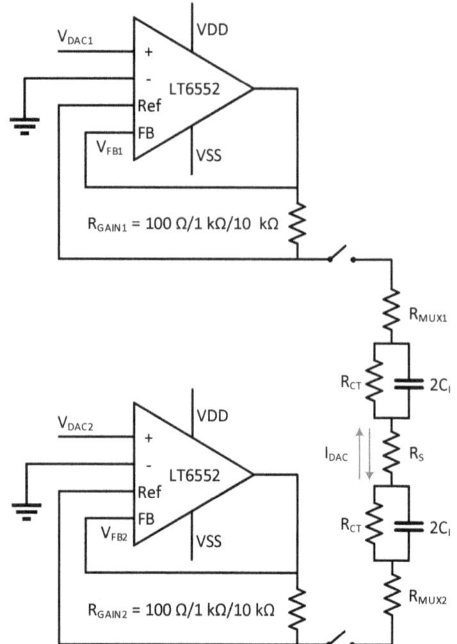

**Figure 2.** Bipolar current source architecture.

### 2.3. Digital Timing Control

Digital timing for the pulse mode had 6 phases (W1, P1, GP, P2, W2, and ST) in one stimulation cycle (Figure 3a). Each phase had a resolution of 1 µs with 16 bits of range (65,536 µs max) and could be programmed independently. The stimulation frequency (cycle repetition rate) was independent to be set through the wait periods (W1 and W2). Stimulation frequencies lower than 15 Hz were implemented with software control using the Python interface. The DAC was configured using SPI to operate in differential mode prior to the start of digital stimulation cycle. During the first pulse (P1), both multiplexers were enabled to allow current to flow through the electrodes, while for the second pulse (P2), the multiplexer addresses were swapped so the polarity of current flowing was reversed. During the shorting time (ST), the grounding switches were enabled to short the electrodes to ground.

Arbitrary waveform mode used a lookup table to update the output current at pre-defined intervals. The lookup tables were written in the FPGA firmware to ensure high timing precision. Each point from the lookup table was written into DAC using SPI at a minimum delay of 2 µs. Note that the multiplexer control and grounding signals could be controlled simultaneously with an arbitrary waveform. Key specifications for MEDUSA are shown in Table 1.

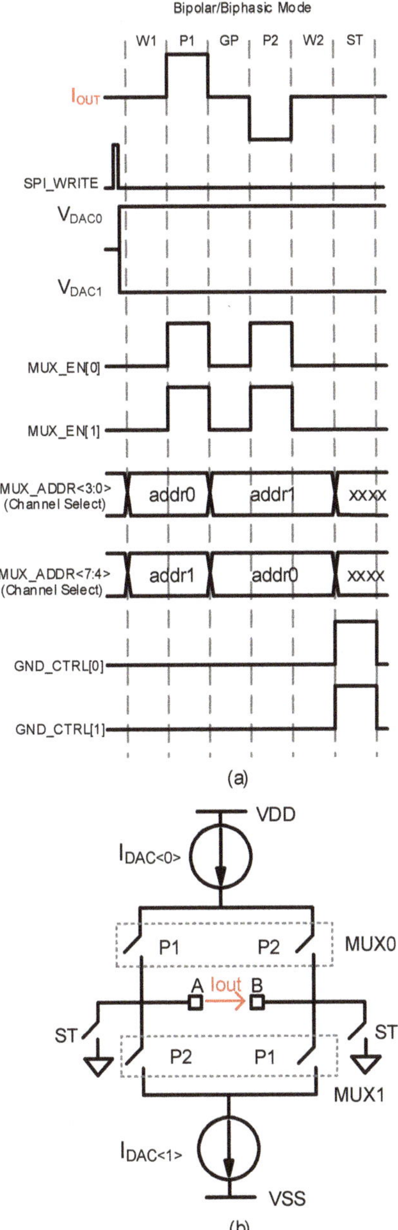



**Figure 3.** (**a**) Example timing for biphasic, bipolar stimulation and (**b**) the resulting stimulation circuit.

**Table 1.** MEDUSA specifications.

| Electrical Specifications | | | |
|---|---|---|---|
| # of Current Sources | | 8 | |
| # of Output Channels | | 16 | |
| Current Resolution | Range 1 (R = 10 kΩ) | 15.3 | |
| | Range 2 (R = 1 kΩ) | 153 | nA |
| | Range 3 (R = 100 Ω) | 1530 | |
| Current Range | Range 1 (R = 10 kΩ) | ±0.5 | |
| | Range 2 (R = 1 kΩ) | ±3 | mA |
| | Range 3 (R = 100 Ω) | ±5 | |
| Compliance | ±5 mA | ±5 | V |
| Timing Specifications | | | |
| Pulse Mode | | | |
| Wait Period (W1) | | | |
| Pulse (P1) | | | |
| Interphase Gap (GP) | | Min: 0 | |
| Pulse (P2) | | Max: 65,535 | µs |
| Wait Period (W2) | | Resolution: 1 | |
| Shorting Period (ST) | | | |
| Rise Time | | 120 | ns |
| Arbitrary Waveform Mode | | | |
| DAC Update Rate | | 2 | µs |

## 3. Results

### 3.1. System Transfer Function

To validate the transfer function of MEDUSA, we measured the output current for all three gain modes ($R_{GAIN}$ = 100 Ω, 1 kΩ, and 10 kΩ) for current ranges of ±5 mA, ±2 mA, and ±250 µA. The output current was measured using a source meter at 0 V. The output current vs. digital DAC code is shown in Figure 4a, where the digital code was stepped into increments of 1000. To verify the linearity and resolution, we swept 20 codes around the zero-crossing and measured an LSB of 15.3 nA with a DNL of 0.13 LSB over that range (Figure 4b). Typically, such small currents are not biologically relevant for neuromodulation systems; however, microelectrode electroplating current levels can be on the order of 100 nA [18]. Note that the input bias current of the amplifier was roughly 10 µA, but this offset was canceled through DAC calibration.

### 3.2. Compliance Voltage

To verify the compliance voltage of the stimulator, we swept the load voltage on the source meter for programmed currents (Figure 5). For anodic stimulation, the current output deviated from its specification by 10% around 3.6 V (VDD − 1.4 V). To ensure 5 V compliance, a VDD of 7 V could be used. Cathodic stimulation, however, was compliant with VSS due to the input common-mode range of the current source amplifier.

### 3.3. Arbitrary Waveform Generation

To demonstrate arbitrary waveform generation, we measured the voltage across a 1 kΩ resistive load to ground during stimulation. Figure 6a shows a conventional biphasic waveform with rectangular pulses. All timing and amplitude parameters were configured by the user through the Python interface. The FPGA firmware implemented the timing diagram as shown in Figure 3a to generate the pulses, while the current amplitude for each phase was independently programmed via SPI. Each state of the waveform had a resolution of 1 µs with 16 bits of programmability. The system could also generate arbitrary current stimulation (Figure 6b). In arbitrary waveform mode, the firmware used a lookup table that could be scaled dynamically through the Python interface. The temporal resolution in arbitrary waveform mode was 2 µs, which was limited by the speed at which the DAC could update over SPI. In this mode of operation, the user could customize the scaling factor of the lookup table for both the period and amplitude of the arbitrary waveform.

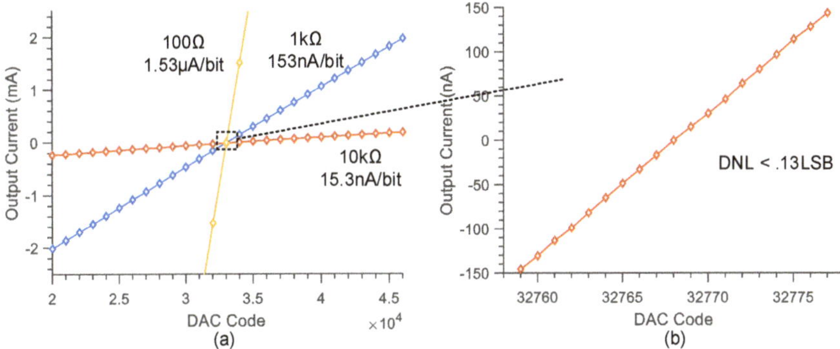


**Figure 4.** (**a**) Measured transfer function for three gain settings ($R_{GAIN}$ = 100 Ω, 1 kΩ, and 10 kΩ). (**b**) Measured linearity around the transition point for high-precision setting (low-gain mode, 10 kΩ).

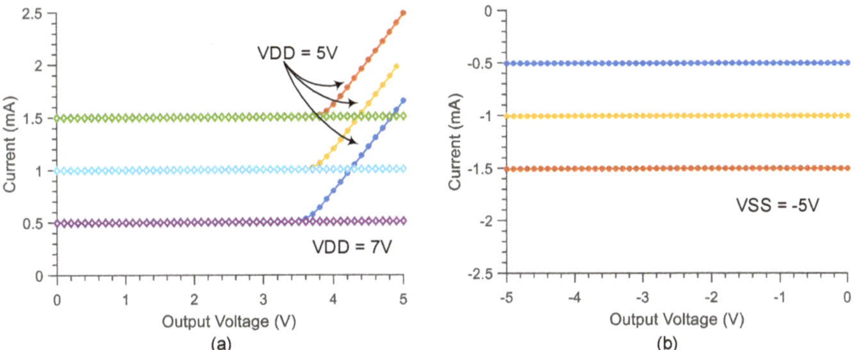


**Figure 5.** Measured voltage compliance for (**a**) anodic and (**b**) cathodic stimulation.

MEDUSA can easily achieve high-frequency stimulation (>1 kHz) without issue. The analog bandwidth was more than 2 MHz, so the stimulation frequency in arbitrary waveform mode was only limited by the DAC update speed. Figure 7a,b shows two biphasic stimulation waveforms at 1 kHz and 10 kHz using two types of stimulation methods discussed above. In Figure 7a, the 1 kHz biphasic stimulation was operating under arbitrary waveform mode using a lookup table update, while in Figure 7b, the 10 kHz stimulation used conventional MUX switch mode with DAC output preset at a certain voltage level.

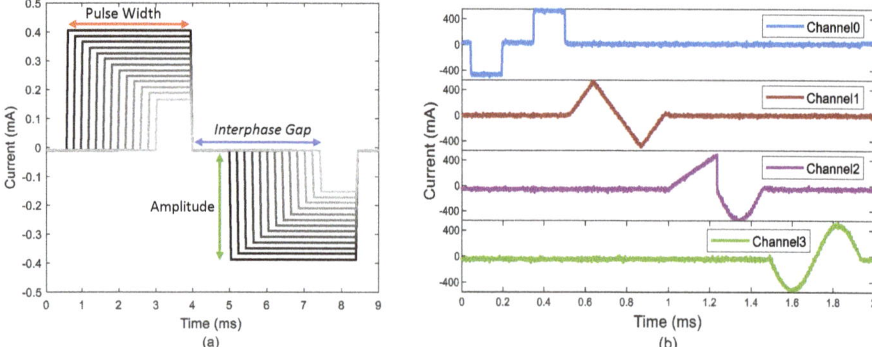



**Figure 6.** (**a**) Pulse width, interphase gap, and amplitude for both positive and negative pulse are customizable and (**b**) arbitrary waveform done using lookup table update.

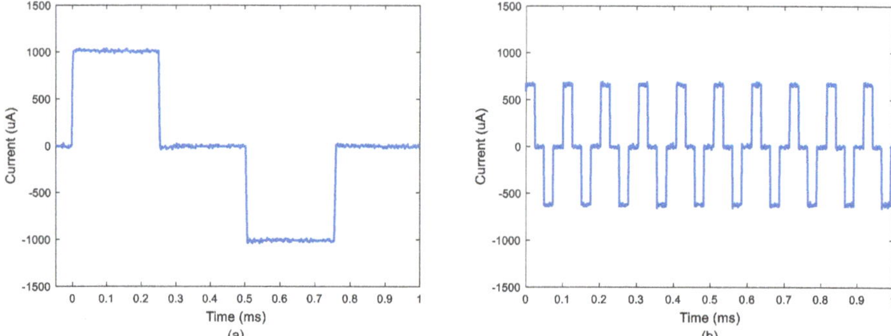



**Figure 7.** Biphasic stim waveform measured across a 1 kΩ resistive load at (**a**) 1 kHz with 1 mA pulse amplitude and (**b**) 10 kHz with amplitude 0.6 mA.

### 3.4. Charge-Balancing In Vitro

To demonstrate charge-balancing using MEDUSA in vitro, we immersed a custom microelectrode array in a saline bath and measured the voltage profile in response to biphasic stimulation. The electrode array was a rigid PCB with 0.5 mm diameter square immersion gold (ENIG) electrodes. A 3D-printed ring defined a well around the electrodes. While these electrodes had poor charge storage capacity relative to conventional stimulation electrodes, such as platinum-iridium, they were an apt test bench for charge-balance due to their nonlinearity during stimulation. As shown in Figure 8a, a balanced biphasic pulse (blue trace, $I_{DAC} = 275$ µA, $P1 = P2 = 26$ µs) generated a significant offset voltage mostly due to appreciable charge leakage during the interphase gap ($GP = 26$ µs). Note that the offset voltage would eventually return to zero since one electrode was always at ground. To compensate for charge leakage during the interphase gap, less charge could be injected in the second phase by lowering its amplitude (red trace) or decreasing its duration. However, a practical implementation requires monitoring the offset after each stimulation sequence. For this reason, we implemented a grounding phase (ST) when electrodes could be shorted to ground, clearing any residual charge on the electrodes. To demonstrate the effect of the shorting switch, we added a 10 µs wait time (W2) after the second pulse before enabling the shorting switch (yellow trace).

To confirm that leakage during the interphase gap was the source of the residual voltage offset, we created a Randles circuit model of the electrodes and simulated it with an ideal stimulator (purple trace, Figure 8a). Figure 8b is the Randles circuit model of the electrode-saline interface, and the biphasic current was injected from the top, illustrated by a current source $I_{DAC}$. The values were determined empirically. During P1, positive current flowed through the electrode-saline interface, and the voltage jumped to $I_{DAC} \cdot R_s$ ($R_s$ is the spread resistance) and started to charge $C_I$. During GP, the stimulation current was zero, and the electrode capacitor discharged. During P2, negative current (reverse polarity) flowed through network and discharged the electrode. If zero charge had been lost during GP, the electrode voltage would return to zero as the stimulation phases were perfectly matched. In this case, the lost charge during GP meant a negative offset voltage remained on the electrode. During ST, a shorting switch was enabled to eliminate the voltage offset.

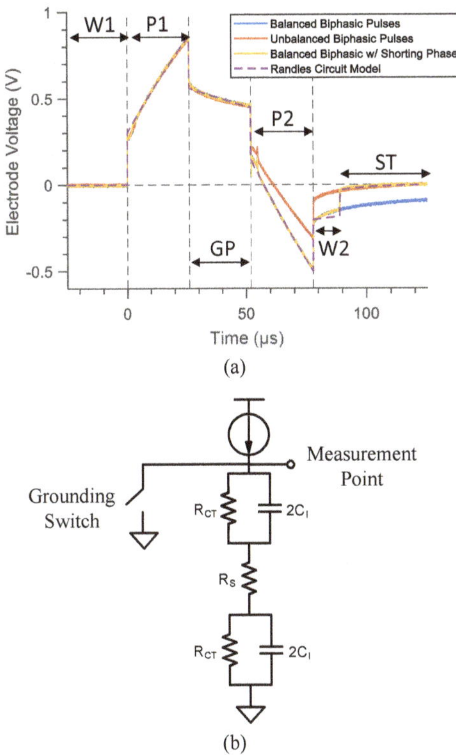


**Figure 8.** (**a**) In vitro measurement results with biphasic simulation pulses and (**b**) Randles circuit model of two series electrodes where $I_{DAC} = 275\ \mu A$, $C_I = 10.4\ nF$, $R_S = 1.2\ k\Omega$, and $R_{CT} = 11.2\ k\Omega$.

### 3.5. Selective Stimulation Protocols

### 3.5.1. Temporal Interference

To demonstrate temporal interference [7], we injected two sine-wave currents (Figure 9a,c) at frequencies of 1 kHz and 1.02 kHz into a resistor mesh in Figure 9, which was used as a simplified electrical model for tissue. The sine-wave current was realized by updating DAC alternately with data from two different lookup tables and then sending currents to two channels simultaneously. This 2% difference of the number of points in these two lookup tables was able to create a 20 Hz frequency offset when one of the fast (100-point) sine-waves was operating at 1 kHz. Furthermore, the frequency and amplitude of sine-wave current could be independently configured to meet certain conditions. The current measured in the center resistor was modulated with an envelope frequency of around 20 Hz (Figure 9b), equal to the offset frequency of source currents.

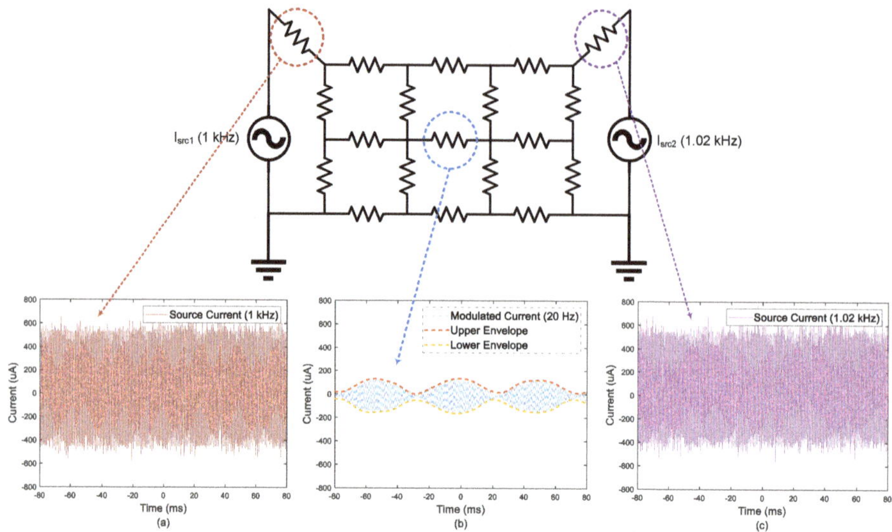


**Figure 9.** (**a**) Modulated current is 20 Hz and (**b**) 1 kHz sine current source and (**c**) 1.02 kHz sine current source.

### 3.5.2. Intersectional Short Pulse

To demonstrate an intersectional short pulse (ISP) [8] stimulation protocol, we measured the voltages across eight resistive loads with 1 kΩ resistance, and they were connected to the corresponding output channels of the stimulation system. The pulse current amplitude was set to be 1 mA and the pulse width to be 10 µs in Figure 10a. The overall charge accumulated at the theoretical focal point showed a linear increase over time (Figure 10b). The ISP function was implemented by updating the DAC via a single lookup table and switching the MUX address incrementally from Address 0 to 7, so the stimulation current pulses were directed to preassigned output channels.

### 3.6. Comparison with State-of-the-Art

To demonstrate MEDUSA's capability for high-frequency neuromodulation, we compared its rise time with a high-end commercial voltage-controlled precision current source (Stanford Research System CS580). For a conventional pulse stimulation, the 10–90% rise times 120 ns and 2.8 µs, respectively, were measured across a 1 kΩ resistive load with a 600 µA current amplitude as shown in Figure 11a. This corresponded to analog bandwidths of 2.9 MHz and 125 kHz. MEDUSA achieved a rapid rise time by switching the multiplexer, allowing for the DAC and current source to settle during the first wait period (W1).

Table 2 shows a comparison of MEDUSA and other commercial or open source neuromodulators. MEDUSA had several key advantages: higher channel count, higher analog bandwidth, and seamless integration with an FPGA for complex waveform generation. Many other current sources required a function generator and a means to input an arbitrary waveform into the function generator, which made scaling to multiple stimulation channels impractical. MEDUSA also had the ability to short the electrodes to ground on command for charge balance to prototype novel in vivo stimulation protocols.

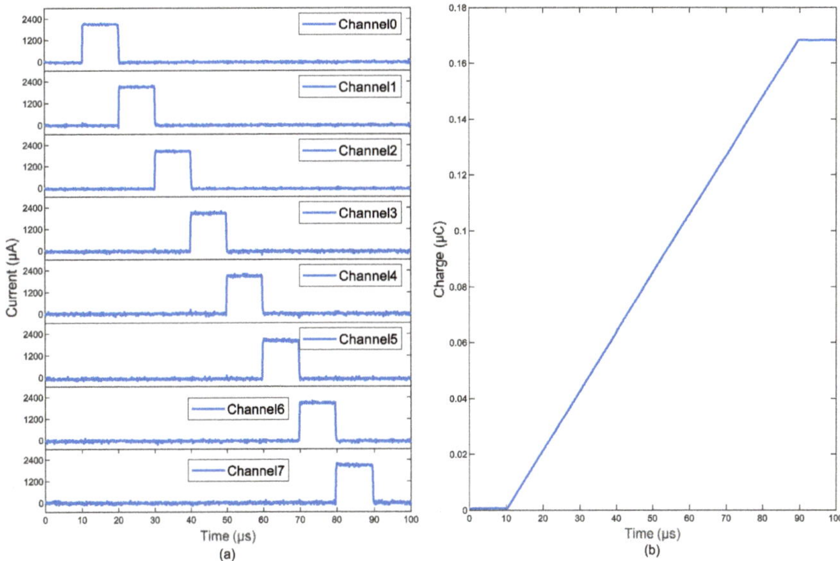

**Figure 10.** (**a**) Intersectional short pulse through eight channels terminated with 1 kΩ load and (**b**) total charge accumulated.

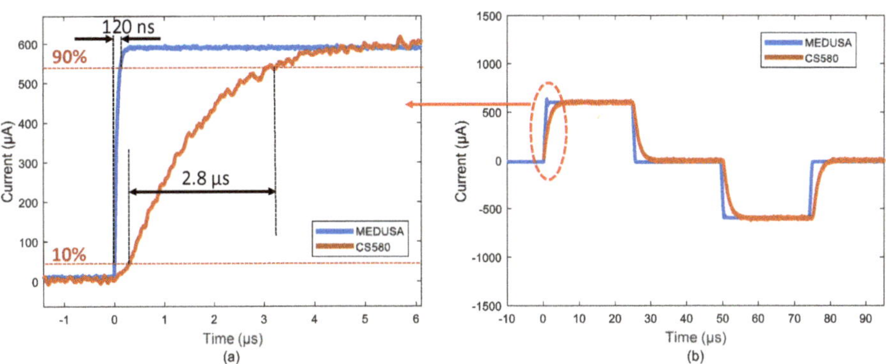

**Figure 11.** The rise time for the MEDUS system and SRS CS580 is 120 ns and 2.8 µs, respectively (**a**), measured across a 1 kΩ resistive load from 10% to 90% of voltage, (**b**) and the 10 kHz stimulation.

Table 2. Overview of neural current stimulators.

| | MEDUSA (This Work) | CS580 Voltage Controlled Current Source | Model 2000 Analog Stimulus Isolator | SYS-A395 Linear Stimulus Isolator | StimJim |
|---|---|---|---|---|---|
| Output Type | Current | Current | Current/Voltage | Current | Current/Voltage |
| # of Current Sources | 8 | 1 | 1 | 1 | 2 |
| # of Output Channels | 16 | 1 | 1 | 1 | 2 |
| Required Input Source | Opal Kelly FPGA | Function Generator | Function Generator | Function Generator | No |
| Source Polarity | Bipolar | Bipolar | Bipolar | Bipolar | Bipolar |
| Analog Bandwidth | 3 MHz | 200 kHz | 40 kHz | 10 kHz | - |
| Maximum Current | $\pm$5 mA | $\pm$100 mA | $\pm$5 mA | $\pm$10 mA | $\pm$1.36 mA |
| Current Resolution | 15.3 nA to 1530 nA | 100 fA to 10 $\mu$A | - | - | 0.1 $\mu$A |
| Rise Time | 120 ns | 2.8 $\mu$s | <10 $\mu$s [a] | 26 $\mu$s [a] | 2 $\mu$s to 6 $\mu$s [b] |
| Active Charge Balance | Yes | - | - | - | Yes |
| Estimated Cost | $200 | $2795 | $1400 | $1869 | $202 |
| Weight | 0.3 lbs | 15 lbs | 2.53 lbs | 4 lbs | - |
| Dimension | $3.6 \times 5.0 \times 0.5$ in$^3$ | $8.3 \times 3.5 \times 13$in$^3$ | $2.5 \times 6.1 \times 6.2$ in$^3$ | $6.5 \times 4 \times 3.5$ in$^3$ | $5.5 \times 4.1 \times 1.37$ in$^3$ |

[a] Given by the datasheet. [b] Estimated based on the plot from the paper.

### 3.7. In Vivo Results

To demonstrate MEDUSA as a research platform, we performed a series of in vivo experiments using a common earthworm (Lumbricus terrestris). While no IACUC protocols were required, we ensured humane treatment by anesthetizing the earthworm using a 10% ethanol solution. An overview of the experimental setup is shown in Figure 12. The earthworm was placed dorsal side up, and electrode pins were inserted through the worm, fixing it in place. The stimulation anode was placed 2 cm caudally of clitellum and separated from the cathode by 1 cm. Recording channel R1 was placed 4 cm below the stimulation cathode. The recording reference (REF) was placed in the center between the stimulus cathode and R1 [19,20].

To build a strength-duration curve (Figure 13), we used monophasic current stimulation and swept pulse width and amplitude to determine the action potential threshold for the median giant nerve (MGN). Stimulation frequency was below 0.5 Hz to ensure that the nerve was not fatigued. The MGN showed a rheobase current of 50 µA and a chronaxy of about 400 µs [21].

To verify a response from the lateral giant nerve (LGN), we used a fixed 100 µs monophasic current pulse and increased amplitude starting at 200 µA. Due to its smaller diameter, the LGN had a higher threshold and slower conduction velocity relative to the MGN. Figure 14a shows stimulus amplitudes below the LGN's response threshold, while Figure 14b shows the response for stimulus amplitudes above the threshold. By comparing the delay between recording channels (R1 to R2), we estimated the conduction velocity of the MGN and LGN to be 20.0 m/s and 7.7 m/s, respectively.

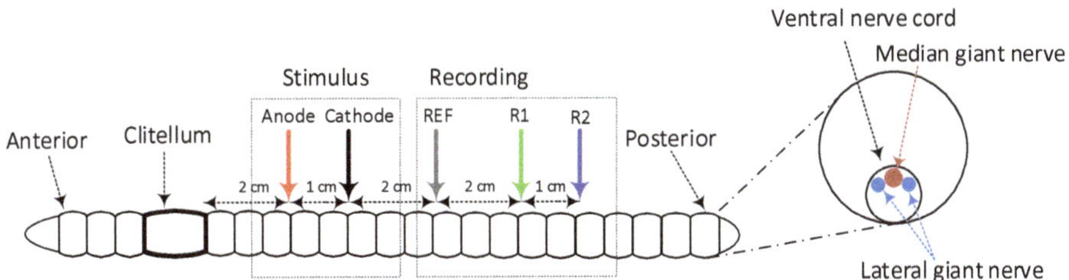


**Figure 12.** Earthworm stimulation experiment setup.

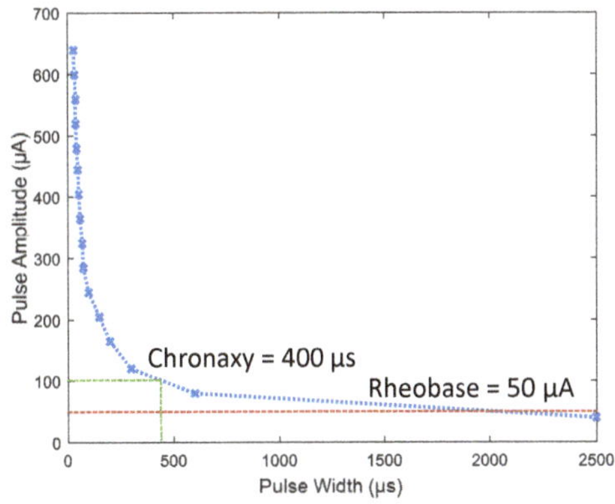


**Figure 13.** Strength-duration curve of earthworm median giant nerve (MGN).

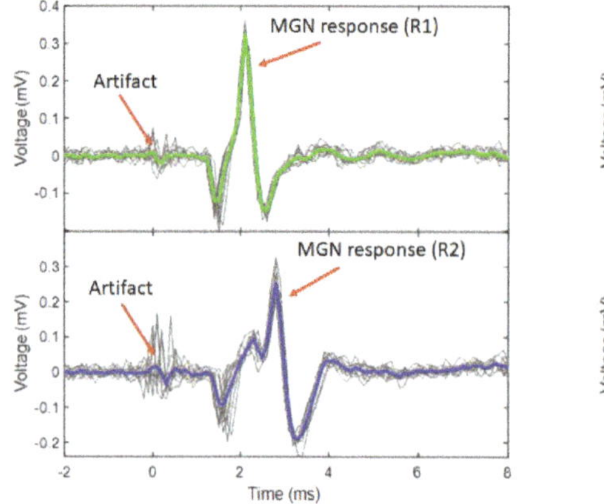
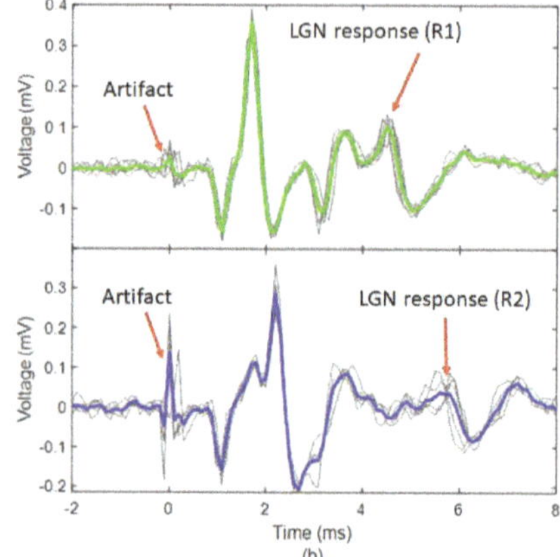


**Figure 14.** Nerve spikes in response to monophasic current stimulation of a 100 μs duration recorded from R1 and R2 with amplitude swept from (**a**) 200 to 650 μA and (**b**) 700 μA to 2 mA. Recordings were made using a commercial recording unit Neuron SpikerBox Pro (Backyard Brains, Ann Arbor, MI, USA).

## 4. Conclusions

We presented MEDUSA, a low-cost and high-performance 16-channel neural stimulation system built from off-the-shelf components. MEDUSA was intended to be a general purpose neuroscience research tool to explore novel, high-frequency, multi-channel stimulation paradigms. Our future work will migrate MEDUSA's core functionality to a wireless system. However, wireless systems are far more constrained in terms of area and power and require more application-specific design implementations compared to general purpose platforms [12,22–26].

MEDUSA achieved a high dynamic range by combing a 16-bit voltage DAC and three orders of magnitude of gain selection in the current source. The system was capable of delivering monophasic and biphasic stimulation pulses with 1 μs temporal resolution, DC currents, and arbitrary waveforms with a temporal resolution of 2 μs. MEDUSA provides researchers with a low-cost, multi-functional neuroscience research tool for electroplating microelectrodes, performing electrical impedance spectroscopy, and examining novel neural stimulation protocols.

**Author Contributions:** Conceptualization, B.C.J.; circuit board design and assembly, F.T.; python software and FPGA firmware, F.T.; manuscript preparation, B.C.J. and F.T. All authors have read and agreed to the published version of the manuscript.

**Funding:** This work was made possible by an Institutional Development Award (IDeA) from the National Institute of General Medical Sciences of the National Institutes of Health under Grant #P20GM103408.

**Conflicts of Interest:** The authors declare no conflicts of interest.

## References

1. Plachta, D.T.T.T.; Gierthmuehlen, M.; Cota, O.; Espinosa, N.; Boeser, F.; Herrera, T.C.; Stieglitz, T.; Zentner, J. Blood pressure control with selective vagal nerve stimulation and minimal side effects. *J. Neural Eng.* **2014**, *11*. [CrossRef] [PubMed]

2. Koopman, F.A.; Chavan, S.S.; Miljko, S.; Grazio, S.; Sokolovic, S.; Schuurman, P.R.; Mehta, A.D.; Levine, Y.A.; Faltys, M.; Zitnik, R.; et al. Vagus nerve stimulation inhibits cytokine production and attenuates disease severity in rheumatoid arthritis. *Proc. Natl. Acad. Sci. USA* **2016**, *113*, 8284–8289. [CrossRef] [PubMed]

3. Chakravarthy, K.; Nava, A.; Christo, P.J.; Williams, K. Review of Recent Advances in Peripheral Nerve Stimulation (PNS). *Curr. Pain Headache Rep.* **2016**, *20*, 60. [CrossRef] [PubMed]

4.  Benabid, A.L.; Pollak, P.; Hoffmann, D.; Gervason, C.; Hommel, M.; Perret, J.E.; de Rougemont, J.; Gao, D.M. Long-term suppression of tremor by chronic stimulation of the ventral intermediate thalamic nucleus. *Lancet* **1991**, *337*, 403–406. [CrossRef]

5.  Merrill, D.R.; Bikson, M.; Jefferys, J.G.R. Electrical stimulation of excitable tissue: Design of efficacious and safe protocols. *J. Neurosci. Methods* **2005**, *141*, 171–198. [CrossRef]

6.  Stanslaski, S.; Afshar, P.; Cong, P.; Giftakis, J.; Stypulkowski, P.; Carlson, D.; Linde, D.; Ullestad, D.; Avestruz, A.T.; Denison, T. Design and validation of a fully implantable, chronic, closed-loop neuromodulation device with concurrent sensing and stimulation. *IEEE Trans. Neural Syst. Rehabil. Eng.* **2012**, *20*, 410–421. [CrossRef]

7.  Grossman, N.; Bono, D.; Dedic, N.; Kodandaramaiah, S.B.; Rudenko, A.; Suk, H.J.; Cassara, A.M.; Neufeld, E.; Kuster, N.; Tsai, L.H.; et al. Noninvasive Deep Brain Stimulation via Temporally Interfering Electric Fields. *Cell* **2017**, *169*, 1029–1041. [CrossRef]

8.  Vöröslakos, M.; Takeuchi, Y.; Brinyiczki, K.; Zombori, T.; Oliva, A.; Fernández-Ruiz, A.; Kozák, G.; Kincses, Z.T.; Iványi, B.; Buzsáki, G.; et al. Direct effects of transcranial electric stimulation on brain circuits in rats and humans. *Nature. Commun.* **2018**, *9*, 1–17. [CrossRef]

9.  Lee, H.M.; Howell, B.; Grill, W.M.; Ghovanloo, M. Stimulation Efficiency with Decaying Exponential Waveforms in a Wirelessly Powered Switched-Capacitor Discharge Stimulation System. *IEEE Trans. Biomed. Eng.* **2018**. [CrossRef]

10. Johnson, B.C.; Gambini, S.; Izyumin, I.; Moin, A.; Zhou, A.; Alexandrov, G.; Santacruz, S.R.; Rabaey, J.M.; Carmena, J.M.; Muller, R. An implantable 700 µW 64-channel neuromodulation IC for simultaneous recording and stimulation with rapid artifact recovery. In Proceedings of the 2017 IEEE Symposium on VLSI Circuits, Kyoto, Japan, 5–8 June 2017; pp. C48–C49. [CrossRef]

11. Taschwer, A.; Butz, N.; Kohler, M.; Rossbach, D.; Manoli, Y. A Charge Balanced Neural Stimulator with 3.3 v to 49 v Supply Compliance and Arbitrary Programmable Current Pulse Shapes. In Proceedings of the 2018 IEEE Biomedical Circuits and Systems Conference, BioCAS, Cleveland, OH, USA, 17–19 October 2018; pp. 1–3. [CrossRef]

12. Zhou, A.; Santacruz, S.R.; Johnson, B.C.; Alexandrov, G.; Moin, A.; Burghardt, F.L.; Rabaey, J.M.; Carmena, J.M.; Muller, R. A wireless and artefact-free 128-channel neuromodulation device for closed-loop stimulation and recording in non-human primates. *Nat. Biomed. Eng.* **2018**, *3*, 15–26. [CrossRef]

13. Xu, J.; Guo, H.; Nguyen, A.T.; Lim, H.; Yang, Z. A bidirectional neuromodulation technology for nerve recording and stimulation. *Micromachines* **2018**, *9*, 538. [CrossRef] [PubMed]

14. FallahRad, M.; Zannou, A.L.; Khadka, N.; Prescott, S.A.; Ratté, S.; Zhang, T.; Esteller, R.; Hershey, B.; Bikson, M. Electrophysiology equipment for reliable study of kHz electrical stimulation. *J. Physiol.* **2019**, *597*, 2131–2137. [CrossRef] [PubMed]

15. Cermak, N.; Wilson, M.A.; Schiller, J.; Newman, J.P. Stimjim: Open source hardware for precise electrical stimulation. *bioRxiv* **2019**, 757716. [CrossRef]

16. Ibba, P.; Falco, A.; Rivadeneyra, A.; Lugli, P. Low-Cost Bio-Impedance Analysis System for the Evaluation of Fruit Ripeness. In Proceedings of the IEEE Sensors, New Delhi, India, 28–31 October 2018; pp. 1–4. [CrossRef]

17. Avery, J.; Dowrick, T.; Faulkner, M.; Goren, N.; Holder, D. A versatile and reproducible multi-frequency electrical impedance tomography system. *Sensors* **2017**, *17*, 280. [CrossRef] [PubMed]

18. Nick, C.; Thielemann, C.; Schlaak, H.F. PEDOT:PSS coated gold nanopillar microelectrodes for neural interfaces. In Proceedings of the 2014 International Conference on Manipulation, Manufacturing and Measurement on the Nanoscale, 3M-NANO, Taipei, Taiwan, 27–31 October 2014; doi:10.1109/3M-NANO.2014.7057309. [CrossRef]

19. Shannon, K.M.; Gage, G.J.; Jankovic, A.; Wilson, W.J.; Marzullo, T.C. Portable conduction velocity experiments using earthworms for the college and high school neuroscience teaching laboratory. *Am. J. Physiol. Adv. Physiol. Educ.* **2014**, *38*, 62–70. [CrossRef]

20. Follmann, R.; Rosa, E.; Stein, W. Dynamics of signal propagation and collision in axons. *Phys. Rev. E Stat. Nonlinear Soft Matter Phys.* **2015**, *92*, 032707. [CrossRef] [PubMed]

21. Kladt, N.; Hanslik, U.; Heinzel, H.G. Teaching basic neurophysiology using intact earthworms. *J. Undergrad. Neurosci. Educ.* **2010**, *9*, 20.

22. Yun, S.; Koh, C.S.; Jeong, J.; Seo, J.; Ahn, S.H.; Choi, G.J.; Shim, S.; Shin, J.; Jung, H.H.; Chang, J.W.; et al. Remote-controlled fully implantable neural stimulator for freely moving small animal. *Electronics* **2019**, *8*, 706. [CrossRef]

23. Johnson, B.C.; Shen, K.; Piech, D.; Ghanbari, M.M.; Li, K.Y.; Neely, R.; Carmena, J.M.; Maharbiz, M.M.; Muller, R. StimDust: A 6.5mm3, wireless ultrasonic peripheral nerve stimulator with 82% peak chip efficiency. In Proceedings of the 2018 IEEE Custom Integrated Circuits Conference, CICC 2018, San Diego, CA, USA, 8–11 April 2018; pp. 1–4. [CrossRef]

24. Lee, B.; Koripalli, M.K.; Jia, Y.; Acosta, J.; Sendi, M.S.E.; Choi, Y.; Ghovanloo, M. An Implantable Peripheral Nerve Recording and Stimulation System for Experiments on Freely Moving Animal Subjects. *Sci. Rep.* **2018**, *8*, 1–12. [CrossRef]

25. Piech, D.K.; Johnson, B.C.; Shen, K.; Ghanbari, M.M.; Li, K.Y.; Neely, R.M.; Kay, J.E.; Carmena, J.M.; Maharbiz, M.M. A wireless millimetre-scale implantable neural stimulator with ultrasonically powered bidirectional communication. *Nat. Biomed. Eng.* **2020**, *4*, 207–222. [CrossRef]

26. Freeman, D.K.; O'Brien, J.M.; Kumar, P.; Daniels, B.; Irion, R.A.; Shraytah, L.; Ingersoll, B.K.; Magyar, A.P.; Czarnecki, A.; Wheeler, J.; et al. A sub-millimeter, inductively powered neural stimulator. *Front. Neurosci.* **2017**, *11*, 659. [CrossRef] [PubMed]

 *electronics*

*Article*

# FPGA-Based Doppler Frequency Estimator for Real-Time Velocimetry

Stefano Ricci *and Valentino Meacci

Information Engineering Department, University of Florence; 50139 Florence, Italy; valentino.meacci@unifi.it
* Correspondence: stefano.ricci@unifi.it

Received: 30 December 2019; Accepted: 6 March 2020; Published: 8 March 2020

**Abstract:** In range-Doppler ultrasound applications, the velocity of a target can be measured by transmitting a mechanical wave, and by evaluating the Doppler shift present on the received echo. Unfortunately, detecting the Doppler shift from the received Doppler spectrum is not a trivial task, and several complex estimators, with different features and performance, have been proposed in the literature for achieving this goal. In several real-time applications, hundreds of thousands of velocity estimates must be produced per second, and not all of the proposed estimators are capable of performing at these high rates. In these challenging conditions, the most widely used approaches are the full centroid frequency estimate or the simple localization of the position of the spectrum peak. The first is more accurate, but the latter features a very quick and straightforward implementation. In this work, we propose an alternative Doppler frequency estimator that merges the advantages of the aforementioned approaches. It exploits the spectrum peak to get an approximate position of the Doppler frequency. Then, centered in this position, a centroid search is applied on a reduced frequency interval to refine the estimate. Doppler simulations are used to compare the accuracy and precision performance of the proposed algorithm with respect to current state of the art approaches. Finally, a Field Programmable Gate Array (FPGA) implementation is proposed that is capable of producing more than 200 k low noise estimates per second, which is suitable for the most demanding real-time applications.

**Keywords:** doppler velocimetry; doppler spectrum; FPGA; centroid estimation

---

## 1. Introduction

Doppler ultrasound is employed in several applications, like the monitoring of industrial processes [1,2], biomedical investigations [3], and non-destructive tests [4]. In pulse-wave Doppler ultrasound, the signal received among subsequent echoes includes information about the displacement—and thus the velocity—of the particles that originated the echo. Ideally, a target investigated with a frequency $F_T$, and moving at constant velocity $v$, produces a single-tone shift whose frequency $f_v$ is given by the well-known Doppler equation:

$$f_v = 2\frac{v}{c}F_T cos(\theta) \tag{1}$$

where $c$ is the sound velocity and $\theta$ is the angle between the target trajectory and the ultrasound propagation [3]. Unfortunately, due to several phenomena like velocity broadening (in a fluid the adjacent particles move with slight different velocities), geometrical broadening (the ultrasound waves sourced by the finite transducer aperture reach the target with slight different angles $\theta$ [5]), or the transit-time broadening (the particle crosses the ultrasound beam in a finite time [6]), the Doppler spectrum is composed by a relatively large bandwidth instead of a single frequency tone.

Several methods have been proposed in literature about how to get the best estimate of the Doppler frequency $f_v$ from the measured Doppler spectrum. Some advanced techniques are based on the detection of the maximum frequency present in the Doppler spectrum [7,8], while others employ mathematical models of the spectrum [9]. The assessment of the spectrum centroid, i.e., the center of mass, is one of the most commonly used approaches:

$$f_v = \frac{\int_{-0.5}^{0.5} x \cdot \mathrm{PSD}(x)dx}{\int_{-0.5}^{0.5} \mathrm{PSD}(x)dx} \tag{2}$$

where PSD is the Power Spectral Density of the Doppler spectrum in the normalized frequency range $-0.5$ to $+0.5$.

In real-time applications, where the Doppler frequency should be evaluated hundreds of thousands of times per second, Equation (2) is sometimes approximated with the frequency that corresponds to the position of the spectrum peak. As shown in the following part of the paper, this approximation allows an easily hardware implementation, but features a lower precision with respect to Equation (2).

In this work, an improved frequency Doppler estimator is presented. The estimator is implemented in the field programmable gate array (FPGA) included in a real-time ultrasound board [10]. The proposed algorithm takes advantage of the characteristics of both the full centroid and the peak estimators: it features the low calculation effort typical of the peak estimator and the reduced variability typical of the centroid estimator. Unlike Equation (2), which is calculated over the full Doppler spectrum range, the proposed algorithm calculates the center of mass in a reduced frequency span. The estimate is carried out with less calculations, which represents a significant advantage for the hardware implementation. A parameter is introduced to set the extension of the frequency range where the centroid is calculated. Ultrasound simulations based on specialized software running in Matlab (The Mathworks, Natick, MA, USA) are employed to optimize this parameter with respect to the features of the Doppler spectrum. Other experiments are carried out to investigate the estimator accuracy and precision, and to show how it compares to state of the art.

The hardware implementation of the estimator is described and evaluated by experiments. A Newtonian fluid flowing in a pipe is investigated through the Doppler board [10], where peak, full centroid, and the proposed estimator are implemented. The throughput of the three estimators is measured and the quality of the implementation is evaluated by comparing the Doppler frequency estimates calculated in hardware to the corresponding estimates obtained in Matlab. Experiments show that the proposed estimator is capable of producing more than 200 k estimates per second with a mathematical noise below $-150$ dB, high accuracy, and precision.

## 2. Background and State-of-the-Art

### 2.1. Doppler Processing Data Path Overview

General details of the data processing employed in Pulse-Wave ultrasound can be found in several papers and books, for example References [3,10–12]. Here a brief resume is reported to help readers with the subject, and to highlight the parts most relevant for the comprehension of the proposed algorithm.

In a pulse-wave Doppler system a burst of ultrasound waves of frequency $F_T$ is transmitted [13] in the fluid to be investigated by high-voltage pulsers [14] every pulse repetition interval (PRI). The ultrasonic burst travels in the medium at velocity $c$, and when it hits a scatterer, part of its energy is reflected towards the transducer. If the scatterer moves at velocity $v$, the echoes from subsequent transmissions are returned from slightly different positions. A phase-shift is thus observed in the received echoes. In particular, the phase rotation of echoes from subsequent PRIs corresponds to the frequency $f_v$ described quantitatively by the Doppler Equation (1).

The echoes received by the transducer are analog conditioned in the system front-end through filtering and amplification, and analog-to-digital (AD) converted at rate $F_c = \frac{1}{T_c}$ (see Figure 1 top-left). From now on, the calculations are typically performed through digital devices, like digital signal processors (DSP) or FPGAs.

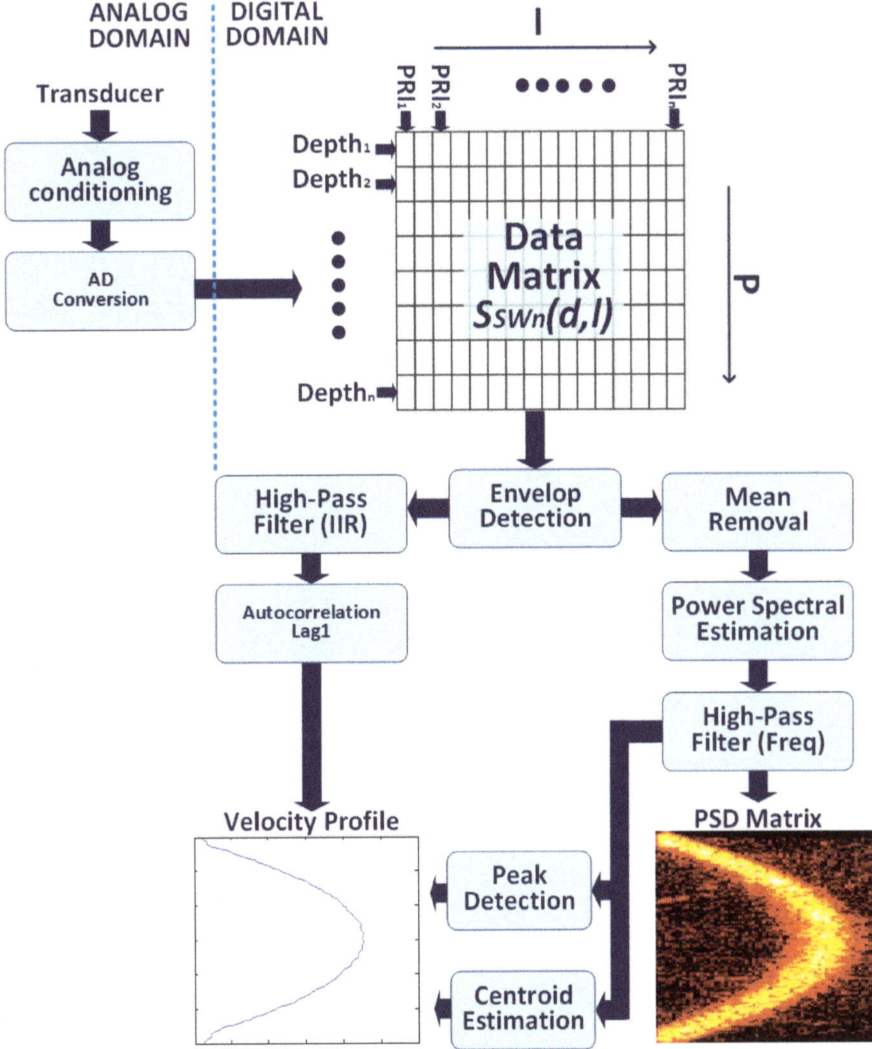



**Figure 1.** Typical Pulsed Wave Doppler processing. Received echoes are analog-to-digital (AD) converted and organized along the columns of a matrix. The signal is demodulated and processed through high-pass filtering and autocorrelation or, alternatively the power spectral density (PSD) is calculated through spectral estimation. From PSD, the velocity is recovered through peak detection or centroid estimation.

The stream of sampled data is subdivided in PRIs sections of length $T_{Pri} = k_M \cdot T_c$, where $k_M$ is the number of samples in each section. This process is formalized by considering the signal sampled at time $t = d \cdot T_c + l \cdot T_{Pri}$, and stored in the bidimensional matrix $s_{SWn}(d,l)$ (Figure 1 top-right). The row index, $d$, is typically named the 'fast-time' index and is in ranges of $0 < d < k_M$, while the column index, $l$, counts the PRIs and is typically named the 'slow-time' index. In this notation, the $d$-index represents the 'depth', i.e., the distance from the transducer [15]. Data sampled at same $d$ behave in the same way, and thus are similarly affected by the phase shift described above, which can be detected by an analysis along the $l$-index.

The signal segments stored along the columns of the matrix are processed for signal envelop detection (Figure 1 middle). Hilbert transform or coherent demodulation is employed [16]. The latter operation, used in this work, is performed by multiplying the signal for the complex sequence $e^{-2\pi i f_T dT_c}$ and by applying a low-pass filter with cut-off frequency $f_{LP}$. This filter affects the axial resolution and is typically tailored to the bandwidth of the transducer.

$$s_{SWDn}(d,l) = s_{SWn}(d,l) \cdot e^{-2\pi i f_T dT_c} \tag{3}$$

The elaboration proceeds with the estimation of the Doppler frequency. This information is encapsulated in the sequence of samples stored in $s_{SWDn}(d,l)$, when considered at the same $d$ index, i.e., at the same distance from the transducer. Different strategies are possible, as sketched by the 2 alternate paths visible in Figure 1 bottom. The left path includes high-pass filtering [17] and autocorrelation (Figure 1 bottom-left). Alternatively (Figure 1 bottom-right), the PSD is calculated [18], followed by peak detection or full centroid estimation. Autocorrelation is specifically employed in applications where the PSD is not necessary. This paper is focused on methods based on PSD. Details of each step are provided in the following sections.

*2.2. Power Spectral Estimation*

The power spectral density distribution of the Doppler spectrum is obtained by processing the demodulated data $s_{SWDn}(d,l)$ by means of spectral estimators. Data are organized in 'packets' defined like:

$$\overline{P}(d,l,L) = [s_{SWDn}(d,l), \ldots \ldots, s_{SWDn}(d,l+i), s_{SWDn}(d,l+L-1)] \tag{4}$$

where L is preferably a power of 2. A simple high-pass filter (at least the mean must be removed), is followed by windowing and Fast Fourier Transform (FFT) [19]. The use of more sophisticated adaptive estimators instead of FFT have been reported in literature as well [15,20,21]; however, more sophisticated adaptive estimators require a higher calculation power that makes the real-time hardware implementation more problematic.

This procedure is applied to all the depths of the demodulated data matrix $s_{SWDn}(d,l)$, for obtaining the corresponding sequences of power spectral density lines:

$$PSD(d,k,F) = \Gamma\{\overline{P}(d,F \cdot L \cdot (1-w), L)\} \tag{5}$$

where $d$ and $k$ are the depth and frequency index, respectively, $\Gamma\{x\}$ is the power spectral estimator employed (e.g., $FFT^2(x)$), F is the frame index that accounts for the matrix sequence, $w$ is the overlap percentage of successive data packets ($0 \leq w < 1$). For example, in case of $L = 128$, $\Gamma\{x\} = FFT^2(x)$, $w = 0.5$, the first 128 demodulated data $s_{SWDn}(d,l)$ ($0 < l < 127$) are processed through a 128-poinf FFT for every depth $d$. The squared output represents the first ($F = 0$) output frame, i.e., the first spectral power density matrix. From now on, every new 64 PRIs, the last $L = 128$ data ($w = 0.5$, i.e., 50% overlap) are processed for generating the next frames ($F = 1,2,3$, etc.).

An approximated high-pass filter (clutter filter) can be applied easily in the frequency domain by removing the lower region of the PSD matrices. The performance of this filter is lower with respect to a full time-domain implementation (e.g., a Finite Impulse Response (FIR) filter). Nevertheless, the performance is sufficient for the application purpose [22] and it can be implemented efficiently in hardware. The final PSD matrix, which is color coded, represents an intuitive picture of the flow profile present in the pipe or the vessel (see, for example Figure 1 bottom-right). The rows report the Doppler shifts, which are proportional to the flow velocity, and the columns report the depths inside the pipe or vessel.

Once the PSD matrix is available, the Doppler frequency is obtained by estimating the centroid frequency or, simply, by taking the spectral peak.

### 2.3. Spectral Peak

The simplest Doppler frequency estimator is the peak estimator. In this case, the Doppler frequency is approximated by the position where the spectrum power features the maximum amplitude:

$$n_p : \text{PSD}(d, n_p) = \text{MAX}\{\text{PSD}(d, l), 0 \le l < L\}. \tag{6}$$

The estimator requires $L - 1$ comparison only. It can be implemented in hardware with a single comparator used serially $L - 1$ times.

### 2.4. Centroid Frequency Estimation

The estimator basically calculates the 'center of mass' of the spectrum (2). In the discrete domain it can be implemented as:

$$n_d = \frac{\sum_{l=1}^{L-1} l \cdot \text{PSD}(d, l)}{\sum_{l=1}^{L-1} \text{PSD}(d, l)} \tag{7}$$

where $\text{PSD}(d, l)$ is the power spectrum density bin of index $l$ ($1 < l < L$) calculated at depth $d$. Please note that the 0-frequency ($l = 0$) is excluded from the calculation. The accuracy of this estimator is high, but requires $L - 1$ multiplications, $2(L - 1)$ additions, and a division for each depth $d$, where L is the number of frequency samples of the PSD.

An approximated high-pass filter can be easily implemented by excluding, in addition to the 0-frequency bin, a wider low-frequency region. For example, Equation (7) can be modified as follows:

$$n_d = \frac{\sum_{l=L_m}^{L-L_m} l \cdot \text{PSD}(d, l)}{\sum_{l=L_m}^{L-L_m} \text{PSD}(d, l)}. \tag{8}$$

In this version the bins $-L_m < l < L_m$ are simply ignored. As anticipated above, the performance of this filter is limited, but remains sufficient for most applications.

## 3. The Proposed Method

### 3.1. Method Basics

In this work we propose a modified Doppler frequency estimator optimized for the hardware implementation in real-time Doppler systems. The method is designed as part of the effort to merge the low calculation requirement of the peak estimator with the accuracy and precision of the centroid estimator. The proposed estimator processes the PSD of the Doppler spectrum estimated with a L-point FFT according to the following steps:

(a) The peak estimator (6) is first applied to obtain $n_p$;
(b) The frequency interval $[n_p - B, n_p + B]$, centered on $n_p$ and of extension $2B + 1$ is considered. More details on B are given below.
(c) The centroid frequency $n_n$, output of the estimator, is estimated in the region located in previous step.

The process is clarified with an example. The 3 panels of Figure 2 report a typical Doppler PSD produced by a scatterer moving at 0.5 m/s investigated by transmitting 3.5 MHz ultrasound bursts. The PSD is obtained by a L = 128 point FFT, and it is represented in the normalized frequency range −0.5 to +0.5. The scatterer movement produces a rough bell-shaped spectrum approximatively located in the region 0.21, 0.35. In this example, the background noise is about 60 dB below the spectrum peak.

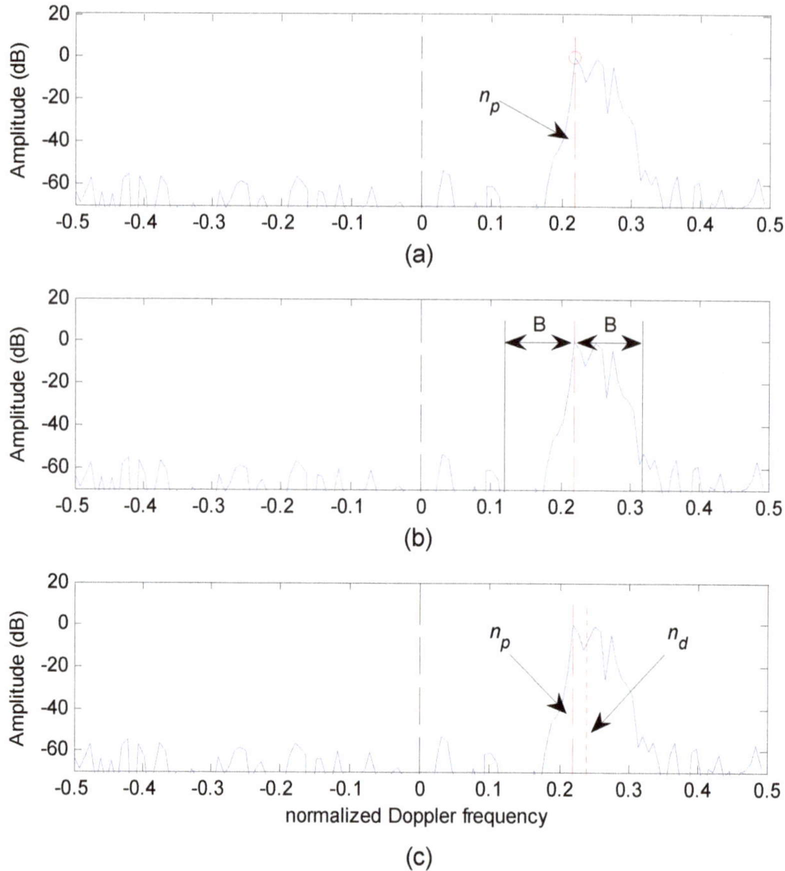

**Figure 2.** The proposed method is applied in 3 steps. (**a**) The spectrum peak (red circle) and the corresponding frequency $n_p$ are detected. (**b**) A frequency interval of extension 2B + 1 is centered around $n_p$. (**c**) The centroid frequency $n_d$ is calculated in the detected interval. Red dashed and dotted segments report $n_p = 0.22$ and $n_d = 0.24$, respectively.

In the first step, the FFT bin corresponding to spectrum peak is located by applying the peak estimator (6). The peak is reported by the red circle in Figure 2a, and the corresponding frequency $n_p = 0.22$ is highlighted by the vertical dashed segment. The detected frequency is considered a first approximation of the Doppler frequency.

In the second step a frequency interval of extension 2B + 1 bins is centered around the $n_p$ frequency. In the example B = 12 bins is chosen, which corresponds to a normalized frequency interval of B/L = 12/128 $\simeq$ 0.1. More details about the selection of the B value are given in the next section. Figure 2b reports the 2B + 1 frequency interval centered in $n_p$. The interval includes approximatively the Doppler bandwidth detectable above the background noise.

In the last steps, the centroid estimator is applied in the frequency region defined in the previous step. The application of the estimator is different with respect to the full centroid calculation (7). In fact, in Equation (7), the whole frequency range is considered, while here the calculation is focalized on the reduced frequency interval centered in the position of the peak. In this case, the centroid estimation is analytically expressed as:

$$n_d = \frac{\sum_{l=n_p-B}^{n_p+B} l \cdot \text{PSD}(d,l)}{\sum_{l=n_p-B}^{n_p+B} \text{PSD}(d,l)}. \tag{9}$$

Figure 2c compares the final Doppler frequency estimate $n_d$ calculated by Equation (9) to the first estimate obtained with the peak estimator $n_p$. The frequencies are reported by dashed and dotted red vertical segments, respectively. In this example, the proposed estimator produces $n_d = 0.24$. This value corrects the first frequency estimate by shifting the result towards the high frequencies. This behavior agrees with the shape and position of the Doppler bandwidth, whose area is visibly biased towards the higher frequency with respect to the $n_p$ estimate.

The proposed estimator requires $L - 1$ comparisons for the detection of the peak in step a), followed by 2B multiplications, 4B summations, and a division for the centroid calculation (see Equation (9)).

### 3.2. The B Parameter

In this section how the performance of the estimator depends on the parameter B is investigated. This study is based on ultrasound simulations carried out with Field II [23,24], a specialized simulator freely available at https://field-ii.dk. Field II, given the position of a set of scattering particles, the transducer characteristics, and the TX pulse, simulates the raw echo data received in each PRI. In this test, a piston transducer transmits in a 45 mm diameter pipe ultrasound bursts composed by 5 cycles at 3.5 MHz and PRI = 0.2 ms. The positions of the scattering particles are updated every PRI to mimic the typical parabolic profile of a Newtonian fluid flowing in a straight pipe. A peak velocity of 0.5 m/s is simulated. White noise generated in Matlab is added to the echo signal produced by the Field II simulator to achieve a Signal-to-Noise Ratios (SNRs) of 0 dB. The signal is further processed as described in Section 2.1 through a 128-point FFT (L = 128) to obtain the PSDs. 20,000 PRIs are generated, corresponding to more than 300 PSD matrices with 50% overlap (w = 0.5, 0 < F ≤ 300). According to the simulation parameters, the nominal normalized Doppler frequency is $F_r = 0.23$.

The proposed estimator (9) is applied to the PSDs at the depth where the velocity reaches its peak of 0.5 m/s. The estimator is applied with the parameter B varying in its full range, i.e., $0 \leq B < 63$. The accuracy of the estimates is evaluated according to the following metrics:

$$\text{Err}_\% = \frac{mean\{F_s(F), F\} - F_r}{F_r} \cdot 100 \tag{10}$$

where $F_s(F)$ is the Doppler frequency detected by the estimator under the test in the PSD of frame F, $F_r$ is the ground truth value obtained by the simulations, $mean\{F_s(F), F\}$ is the average of $F_s(F)$ calculated over the available frames. The precision was evaluated through the Coefficient of Variability (CV%):

$$\text{CV}_\% = \frac{std\{F_s(F), F\} - F_r}{F_r} \cdot 100 \tag{11}$$

where $std\{F_s(F), F\}$ is the standard deviation of $F_s(F)$ calculated in the available frames.

Figure 3 shows the results of the study. The top panel (a) reports the mean spectrum (blue curve), obtained by averaging the 300 available frames. An example of a single spectrum (red dashed curve) is superimposed. The vertical black segment reports the nominal Doppler frequency $F_r = 0.23$. Figure 3b,c report, respectively, the mean error and the coefficient of variability calculated by Equation (10) and Equation (11). Both graphs present a similar trend: the worst performance is observed for low values of B. Performance improves with higher values of B until B reaches 12 (vertical red dashed segments). From now on, performance is stable at its maximum value. The frequency interval corresponding to B = 12 is highlighted in Figure 3a by the vertical dashed segments. The interval includes the Doppler frequencies detectable above the background noise.

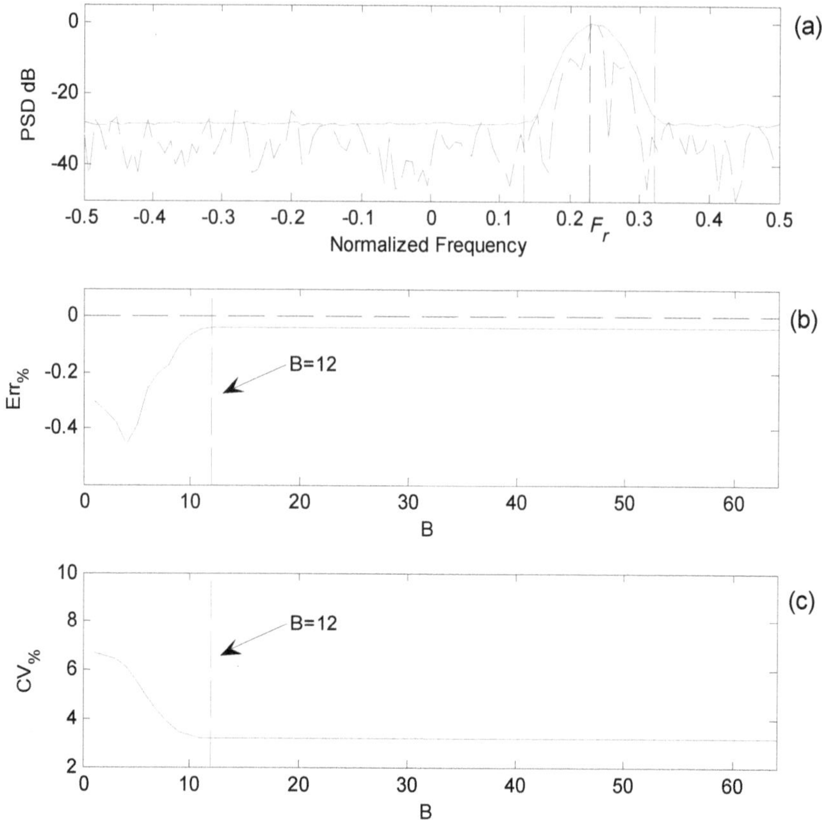

**Figure 3.** Performance of the proposed estimator at different values of B. (**a**) Mean Doppler spectrum (blue curve) and a single spectrum (red dashed curve) produced by a flow signal with a nominal normalized Doppler frequency $F_r$. (**b**) Accuracy achieved for B in the range 0–63. (**c**) Precision achieved for B in the range 0–63. Red dashed segments report the position for B = 12.

This study suggests that the proposed algorithm produces the best performance when B corresponds to a frequency region that includes all of the Doppler frequencies present in the measured spectrum. Higher B values do not produce improvements. Since the calculation effort scales with B, the optimal choice of B is the lowest value that grants the best accuracy and precision. In this example, it corresponds to B = 12. Although this study is based on a specific example, the simulated Doppler spectrum is representative of a wide range of practical conditions. In conclusion, a value of B in the range 10–20 covers most of the practical cases.

*3.3. Comparison of Proposed Method to Standard Methods*

In this section, the proposed method is compared to the standard approaches employed in real-time applications, i.e., the peak and the full centroid estimator.

For this study, the ultrasound signals generated by Field II for the previous study have been used. White noise was added to achieve 6 different Signal-to-Noise Ratios (SNRs) levels, between −20 dB and +20 dB (see leftmost column in Table 1). The signals were processed as described in the previous sections to obtain the PSDs. These were further processed with the peak estimator (6), full centroid estimator (7), and the proposed estimator (9) with B = 12. The achieved accuracy and precision were evaluated with the metric (10) and (11), respectively. Results are listed in Table 1.

**Table 1.** Accuracy and precision.

| SNR | Peak Estimator | | Full Centroid Estimator | | Proposed | |
|---|---|---|---|---|---|---|
| | Err$_\%$ | CV$_\%$ | Err$_\%$ | CV$_\%$ | Err$_\%$ | CV$_\%$ |
| −20 dB | −8.74% | 51.1% | −68.3% | 63.3% | −8.8% | 50.7% |
| −15 dB | −0.38% | 7.3% | −29.67% | 31.9% | −0.11% | 4.4% |
| −10 dB | −0.21% | 6.9% | −2.38% | 6.7% | −0.15% | 3.2% |
| 0 dB | −0.13% | 6.8% | −0.04% | 3.2% | −0.06% | 3.2% |
| 10 dB | −0.17% | 6.8% | −0.04% | 3.2% | −0.04% | 3.2% |
| 20 dB | −0.13% | 6.7% | −0.04% | 3.2% | −0.04% | 3.2% |

When the SNR is below −20 dB, all the tested algorithms fail (first row of Table 1). With SNR = −15 dB (second row), the centroid estimator is still not capable of producing reliable estimates, but peak and proposed estimators start to issue reasonable results. As long as the SNR increases (3rd and following rows of Table 1), the performance improves. In particular, peak estimator produces good results starting from −10 dB (Err$_\%$ $\simeq$ 0.1%–0.2% and CV$_\%$ $\simeq$ 6.7%–6.9%); full centroid estimator features Err$_\%$ = 0.04% and CV$_\%$ $\simeq$ 3.2% from 0 dB. The proposed estimator presents the best performance. In fact, it features a low error and high precision starting from −15 dB, and its precision is always better compared to the peak estimator.

These results are explained by the features of the typical ultrasound signal. The ultrasound signal is affected by the speckle noise [25], which produces temporal variation, which is visible in the example of Figure 2. This noise varies quickly in time, and successive spectra are affected by a completely different noise shape. The peak estimator is particularly prone to this noise, as confirmed by its relatively high variability (CV$_\%$ > 6.7% in Table 1), which is also present with a high SNR. On the other hand, the full centroid and the proposed estimators, which employ a weighted mean over a frequency region, are less sensitive to random variation of the spectra (CV$_\%$ $\simeq$ 3.2% in Table 1).

Peak frequency estimator is affected by another limitation: its output values are limited to the exact positions of the FFT bins, i.e., they are quantized to L values over the whole spectrum range. In other words, the frequency resolution is 1/(PRI·L) only. This limitation does not apply to full centroid or the proposed estimators.

Table 2 reports the number of comparisons, additions, multiplications, and divisions required by each estimator, while the operations required by the other processing steps visible in Figure 1 (e.g., the FFT), are excluded here. The operations are reported as a function of L and B and for the typical values of L = 128 and B = 12 (left and right of the corresponding column). The weight (second column in Table 2) represents a heuristic value that accounts for the complexity of the operations in a generic hardware implementation. Comparisons and additions are considered for weight 1 because in FPGA, its hardware is quite simple. Multiplications require a more complex hardware, so their weight is set to 2. Divisions are the most complex, thus the weight is set to 16. A total effort is then estimated by summing each contribution calculated by the product between the number of operations and the corresponding weight. The result is reported in the last row of Table 2. The method that requires less hardware effort is the peak estimator, but this efficiency is achieved at the expanse of precision (see Table 1). Proposed algorithm features an optimal tradeoff: it presents better accuracy and precision than the full centroid estimator, and requires half of its hardware effort (239/524 = 45%).

**Table 2.** Calculation power for the estimators.

| Operations | Weight | Peak Estimator L = 128 | | Full Centroid Estimator L = 128 | | Proposed Estimator L = 128 B = 12 | |
|---|---|---|---|---|---|---|---|
| Comparisons | 1 | L−1 | 127 | - | - | L−1 | 127 |
| Additions | 1 | - | - | 2(L−1) | 254 | 4B | 48 |
| Multiplications | 2 | - | - | L−1 | 127 | 2B | 24 |
| Divisions | 16 | - | - | 1 | 1 | 1 | 1 |
| Total Effort | | | 127 | | 524 | | 239 |

## 4. Circuit Architecture

The circuit was implemented in a FPGA of the Cyclone III family (Altera-Intel, San Jose, CA, USA). The FPGA is part of a complete ultrasound system designed for the investigation of fluids in industrial applications. More detail of the system and of the implementation of the full processing chain can be found in Reference [10], while this description is limited to the sections of interest for the frequency estimators. The circuit is sketched in Figure 4. A Nios II soft processor (IP from Altera-Intel), manages the sequence of operations and programs the processing modules by setting their registers. The section delimited by the dashed rectangle at the top left of the figure produces the PSDs. The modules delimited by the dashed rectangle located on the Figure bottom implement the frequency estimator. A dual port (DP) memory is used to transfer the PSD between the sections.

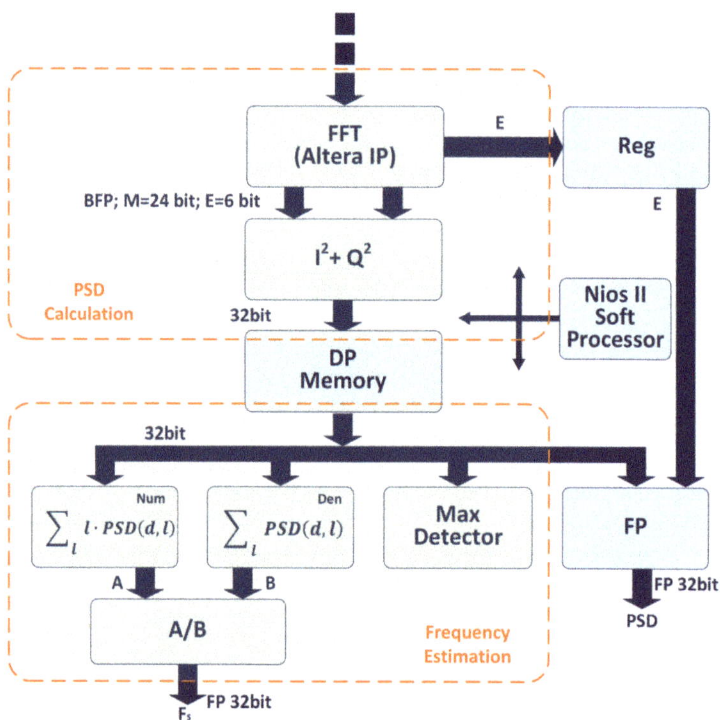


**Figure 4.** Data processing path implemented in the FPGA. The power of FFT output is calculated ($I^2 + Q^2$) and the resulting PSD is saved in a DP memory. Its maximum is detected (Max detector), then the numerator (Num) and denominator (Den) of the centroid formula are calculated. The results are divided, converted in floating point (A/B), and downloaded to host together with the data in memory. A Nios II soft processor supervises the operations.

Data coming from the demodulator are processed through a 128-point FFT working on complex data (I,Q). The FFT is implemented by using the Altera-Intel Intellectual Property (IP) [26], configured with 24 + 24 bit input with "burst with quadruple output" architecture. Once its input buffer is loaded, the IP starts the FFT processing and produces the complex results in the output buffer. The 128-sample result is produced at 24 + 24 bit of mantissa and 6 bit for the exponent. The exponent applies to all the data block (block floating point representation), and is saved in a register for later use. Data are read from the FFT module and moved in the $I^2 + Q^2$ block (see Figure 4), where real and imaginary parts are squared and summed at 48 bit for obtaining the PSD. The 128 PSD samples are reduced to 32 bit and saved in the DP memory. When the 128 data block is completely moved in the memory, the FFT output buffer is empty, and the FFT starts loading and processing a new data block. A floating point converter (FP) is employed to output the PSD in floating point format.

The elaboration proceeds in the second section of the circuit with the estimation of the Doppler frequency from the PSD line present in the DP memory. The electronics implemented in this section is suitable for processing PSD data according to the 3 methods: the peak estimator (6), full centroid estimator (7), and proposed estimator (9). The modules for the Doppler frequency calculation include: the Max Detector, employed to locate the position of the PSD peak; the Num and Den, used to calculate the numerator and denominator of Equation (7) or (9); and a divisor (A/B) for the calculation of the quotient present in Equation (7) or (9). The Nios II processor switches among the 3 different estimators and tunes the parameter B of the proposed estimator by setting the processing module registers: the FPGA does not need to be reconfigured.

In order to process PSD data according to the proposed estimator, the Nios II first activates the Max Detector block, detailed in Figure 5. The Nios II sets the address generator counter (CNT) to scan the full address range (i.e., 0–127) of the DP memory, and zeros the *Max* register. PSD samples are sequentially read from the DP memory, and the comparator (CMP) triggers the enable signal when a value higher than that already stored in the *Max* register is found. The trigger activates the 2 registers that store the new value and its address, respectively. After 128 cycles, the whole DP memory is scanned, and the Max Detector block outputs $n_p$, i.e., the address (position in the spectrum) of the highest PSD sample. In the next step, the Nios II soft processor activates Num and Den modules to calculate numerator and denominator of (9). These modules are detailed in Figure 6. Nios II calculates $n_p - B$ and $n_p + B$, and sets these values in the address generator (CNT) registers to scan the corresponding DP memory span. According to this architecture, the B value can be easily changed at the run time. The accumulation registers A and B are zeroes and the memory is read. As long as the PSD samples are read from the memory, they are accumulated in the B register (see the denominator of Equation (9)). Simultaneously, they are multiplied to the address and the result is accumulated in the A register (see the numerator of Equation (9)). The operation terminates after the programmed memory span is read, i.e., after $n_p + B - (n_p - B) + 1 = 2B + 1$ cycles. In the last step, the Nios II processor starts the A/B module. This module, composed by IPs of the Intel/Altera library, calculates the ratio and converts the result in 32 bit floating-point format. This ratio represents the final frequency estimate calculated according to Equation (9).

1. When data is processed according to Peak estimator, the Nios II activates the Max Detector only to get $n_p$. When data is processed through full centroid estimator (7), the Nios II runs Num and Den modules on the whole PSD frequency range, followed by the A/B module.

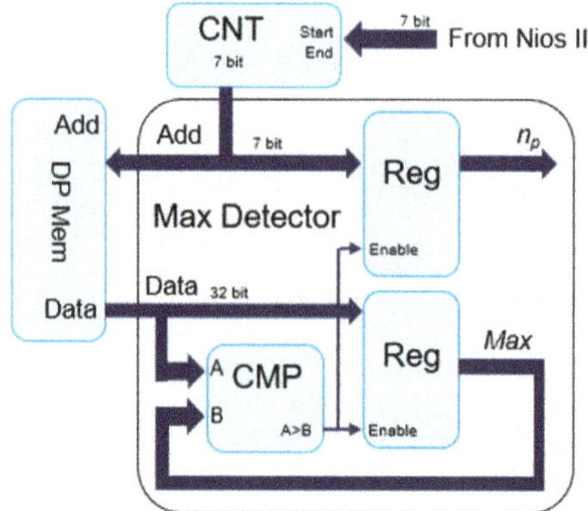



**Figure 5.** Details of the implementation of the Max Detector.

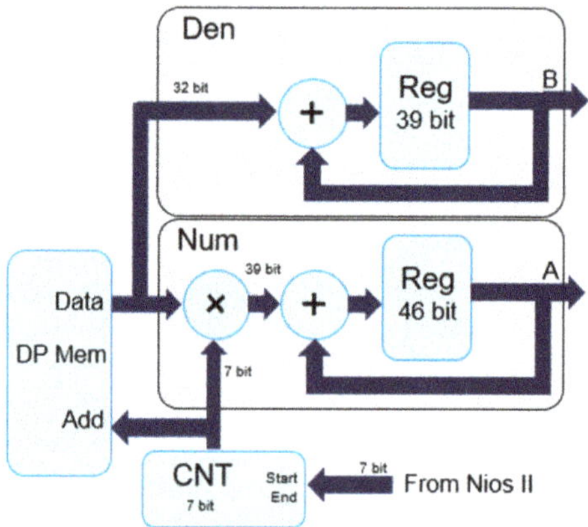



**Figure 6.** Details of the implementation of the Num and Den modules.

## 5. Experiments and Results

The architecture presented in Figure 4 was implemented in the Cyclone III FPGA of the ultrasound board [10]. With the exception of FFT, the DP memory, and the A/B module, which are built with Altera-Intel IPs, all the other blocks are coded for the specific applications directly in VHDL. First, the modules were implemented separately for investigating the latency, maximum clock frequency, and the required FPGA resources. Then, they were connected according to the architecture of Figure 4 and they were employed to process data in real-time in Doppler acquisition. In this last configuration, the throughput and the mathematical noise were analyzed.

### 5.1. Modules Latency

The number of clock cycles needed by each module for processing a 128-sample data set is reported in Table 3. The FFT needs 294 cycles to load inputs, process data, and store results in its output buffer. $I^2 + Q^2$ and FP modules take 135 clock cycles for processing the 128-sample set, corresponding to 1 sample per clock in addition to 7 cycles needed for the initialization and the filling of their pipeline. Similarly, Max Detector processes 1 sample per clock in addition to 2 extra cycles, and Num and Den processes 1 sample per clock in addition to 5 extra cycles. The divisor performs the ratio and floating-point conversion in 16 cycles. First rows of Table 3 lists the number of clock cycles required by the FFT, $I^2 + Q^2$, and FP modules for the calculation of the PSD. The following rows of the Table compares the number of clock cycles required by the Max Detector, Num, Den, A/B modules when implementing the proposed (9), peak (6) and centroid (7) estimators, respectively.

**Table 3.** Latency in clock cycles.

| PSD Calculation (L = 128) | | |
|---|---|---|
| **Block** | **Parameter** | **Latency (Clock Cycles)** |
| FFT | $T_{FFT}$ | 294 |
| $I^2+Q^2$ | $T_{IQ}$ | 135 |
| FP | $T_{FP}$ | 135 |
| **Doppler Frequency Calculation** | | | | |
| **Block** | **Parameter** | **Peak Estimator L = 128** | **Full Centroid Estimator L = 128** | **Proposed est. L = 128; B = 12** |
| Max Det. | $T_M$ | 130 | Not used | 130 |
| Num | $T_{ND}$ | Not used | 131 | 29 |
| Den | $T_{ND}$ | Not used | 131 | 29 |
| A/B | $T_D$ | Not used | 16 | 16 |

*5.2. FPGA Resources and Maximum Clock Frequency*

The resources required by each module are listed in Table 4 together with the maximum clock frequency the module can work at. FFT is the module with the lowest maximum frequency. This is expected, since FFT is the most complex block. Apart from the memory, which is a hardware IP, the highest frequency is achieved by the Max Detector.

**Table 4.** Resource utilization and frequency of each processing module in a Cyclone III FPGA.

| Block | LEs | DSPs | Memory Bits | Max Clock Freq. |
|---|---|---|---|---|
| FFT | 10,308 | 24 | 40,900 | 105 MHz |
| $I^2 + Q^2$ | 400 | 14 | 0 | 110 MHz |
| DP Mem | 0 | 0 | 4096 | 210 MHz |
| Max Det. | 70 | 0 | 0 | 150 MHz |
| Num | 600 | 0 | 0 | 120 MHz |
| Den | 600 | 0 | 0 | 120 MHz |
| A/B | 230 | 0 | 0 | 110 MHz |
| FP | 300 | 0 | 0 | 120 MHz |

*5.3. Data Throughput*

The modules were connected together like in Figure 4 and fed by a 100 MHz clock. For each module, a "Busy" signal was generated. Busy is high when the corresponding module is processing data, while it is low when the module is ready to accept a new data set. The processed data are available on the high-to-low transition for each module. A "DataValid" signal was added as well. It is asserted for a cycle when the final Doppler frequency estimate is ready. These signals were monitored by connectoing the board [10] to the Altera-Intel in-line debug tools through the Joint Test Action Group (JTAG) channel. The results of this analysis are reported in Figure 7 for the 3 frequency estimators. The horizontal axes report the clock cycles (see top scale in each panel); the origin is arbitrary set to the FFT start.

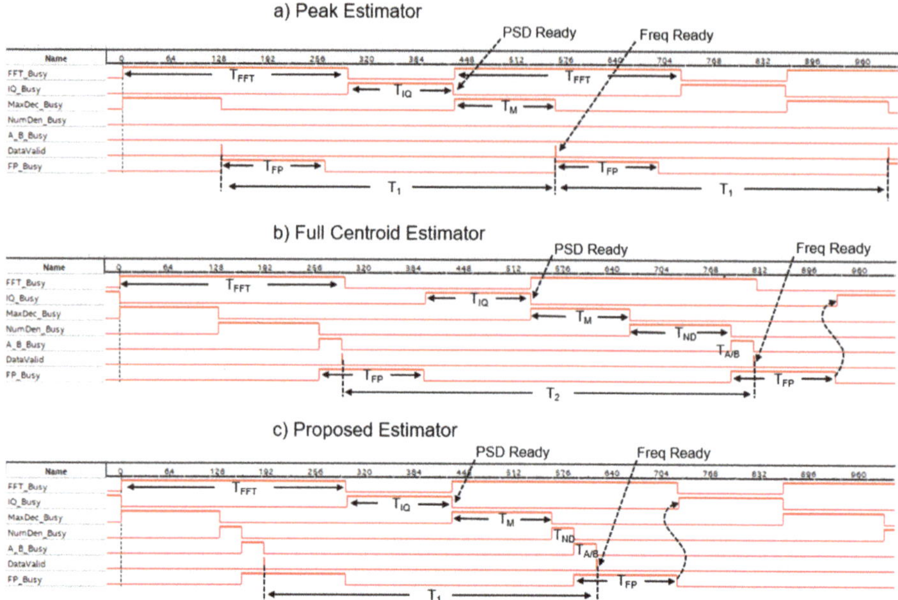

**Figure 7.** Throughput of the processing chain of Figure 4 for (**a**) the Peak, (**b**) Full centroid and (**c**) proposed frequency estimators. The "busy" signals represent status of the corresponding processing block. The DataValid is asserted when the Doppler frequency estimate is ready.

Figure 7a refers to the Peak estimator. The first 2 rows show the working cycles of FFT and $I^2+Q^2$ modules while they calculated the PSD. The FFT restarted its calculation cycle when its output buffer was empty, a condition that occurred when the $I^2 + Q^2$ block had moved the data to DP memory. PSDs were calculated and stored in the DP memory at the maximum throughput $T_1$ allowed by the modules:

$$T_1 = T_{FFT} + T_{IQ} = 294 + 135 = 429 \text{ clock cycles.} \tag{12}$$

Every time a PSD was ready, the Max Detector produced the results (Freq ready) after $T_M = 130$ cycles, as confirmed by the assertion of the DataValid signal (6th row). The Max detector was much faster than FFT and $I^2 + Q^2$ modules (429 against 130 cycles, see Table 3) and stalled for most of the time. The throughput was limited by the calculation of the PSD: a new frequency estimate was produced every $T_1$ cycles.

Figure 7b reports the signals for the full centroid estimator. In this case, the PSD calculation was slower with respect to Equation (12). In fact, the $I^2 + Q^2$ module did not start immediately at the FFT end not to overwrite data in the DP memory that are used by the FP module (see curved arrow). The throughput was here limited by the estimator operations. The clock cycles needed for producing every new frequency estimate were:

$$T_2 = T_M + T_{ND} + T_{FP} + T_{IQ} = 130 + 131 + 135 + 135 = 531. \tag{13}$$

The module A/B worked in parallel to FP, so it did not affect the timings to the point of achieving $T_D < T_{FP}$.

Figure 7c reports the signals for the proposed estimator. In this case, $T_{ND}$ was much lower compared to the previous case, and the FP module completed its operations near the end of the FFT. Thus, the DP memory was free and $I^2 + Q^2$ started the calculations as soon as FFT ended, like in

Figure 7a. In other words, the bottleneck was not the estimator, but the calculation of the PSD, as in the case of the Peak estimator, and Equation (12) applies again.

Table 5 summarizes the throughput of the architecture of Figure 4 for the 3 estimators considered here. When the Peak estimator is employed, the throughput is limited by the calculations of the PSDs, which depends on the performance of FFT; when the Full centroid estimator is used it represents the bottleneck. The proposed estimator is faster than the Full centroid, and the performance is limited by the PSD calculation, like for the Peak estimator.

**Table 5.** Throughput of the frequency estimation for L = 128.

| Estimator | T (Clock Cycles) | Throughput Estimates/s |
| --- | --- | --- |
| Proposed B = 12 | 429 | 232 k |
| Full centroid | 531 | 189 k |
| Peak estimator | 429 | 232 k |

*5.4. Mathematical Noise*

The ultrasound board was used to investigate a flow in a pipe of 40 mm diameter. In the pipe flowed a Newtonian suspension composed by 10 μm plastic spheres dispersed in demineralized and degassed water. The plastic particles were responsible for ultrasound scattering [27]. A focused piston transducer transmitted short bursts at 3.5 MHz every 500 μs. Received echoes were processed onboard in real-time through the 3 estimators considered in this paper. As detailed in Section 4, it was possible to switch among the different estimators by changing the operation sequence in the Nios II processor without reconfiguring the FPGA. The proposed estimator, in particular, was set with B = 12.

Raw data from the converters, calculated PSDs, and frequency profiles calculated by the proposed estimator where saved onboard and then downloaded to a PC for postprocessing and further analysis in Matlab.

An example of PSD matrix calculated by the board is reported in the left panel of Figure 8. The spectral power density, coded in colors in 40 dB dynamics, clearly shows the parabolic profile developed by the flow. The background noise is visible at about −30 dB. Reference PSDs were calculated in Matlab by using re-processing in its native 64-bit floating point representation the data acquired by the onboard AD converters. Onboard- and Matlab-calculated PSD were compared for the evaluation of the mathematical noise introduced by the FPGA calculations. The SNR, evaluated by computing the ratio between the power of the Matlab PSDs and the difference between Matlab and FPGA PSDs, was higher than 130 dB.

The middle panel of Figure 8 shows the frequency profiles calculated onboard with the proposed algorithm from the PSD matrix visible on the left. The frequency profiles calculated on board with the proposed estimator were compared to those calculated in Matlab starting from the same PSDs. The SNR, quantified in the same way as for the PSDs, resulted in a higher 150 dB. The right panel of Figure 8 shows, for example, the difference between the frequency profile reported in the middle panel and the corresponding profile calculated in Matlab.

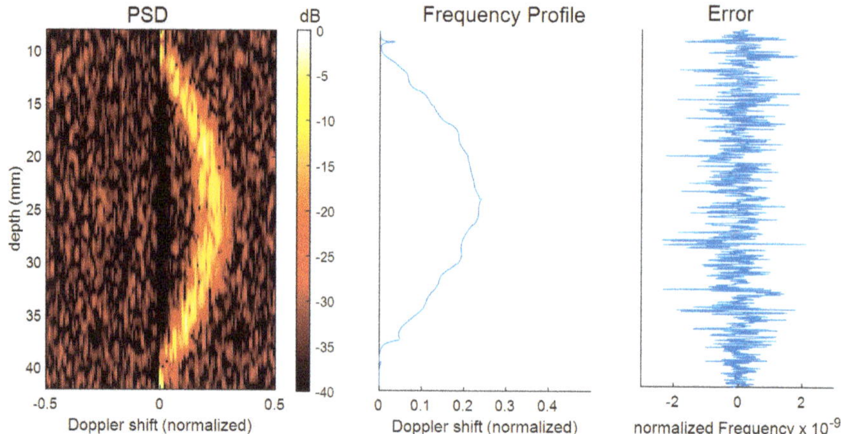

**Figure 8.** A 0.5 m/s flow was investigated with a focused transducer transmitting short 6 MHz frequency bursts. An example of power spectral density (PSD) matrix (left), frequency profile (center) and its error (right), calculated in the FPGA are reported.

## 6. Discussion and Conclusions

In this work, an efficient estimator for the detection of Doppler frequency from spectra has been presented. Its FPGA implementation was reported as well. The estimator exploits and merges the advantages of the peak and full centroid estimators, typically employed in real-time Doppler applications. The proposed estimator features an accuracy similar to or higher than that achieved by the full centroid estimator, and in addition allows a 50% (with B = 12) saving in calculation power (see Table 2).

The estimator employs the parameter B. Simulations allowed us to estimate B = 12 similarly to the optimal choice, at least in the presented case of study. Theoretically this value should be tailored to the extension of the "bell" of the spectrum. However, its exact value is not critical and a value between 10 and 20 is expected to fit most of the practical applications. Moreover, in the presented FPGA implementation, B can be changed at run-time, as well as among subsequent PSDs, if required. In a critical application, the Nios II can be programmed to calculate statistics like those that presented in Figure 3 to detect optimal B values.

Literature reports errors around 4%–10% for ultrasound measurements [28]. The error reported in this paper (Err$_\%$ < 1% in Table 1) is not in conflict with literature, since it was obtained in simulation, where most of the uncertainties present in real experiments are avoided [29]. On the other hand, simulation results confirm that the proposed estimator achieves a better accuracy with respect to peak estimator.

The presented FPGA architecture is suitable for the calculation of the 3 estimators without reconfiguration, with low resource utilization, and with a mathematical SNR higher than 150 dB. Moreover, with a 100 MHz clock frequency, more than 200 k PSD/s are calculated. This makes the estimator suitable for a wide range of demanding applications, like biomedical real-time blood flow investigations based on vector Doppler and plane waves [30], or analysis of industrial flow in large pipes where thousands of depths per frame are used [4].

**Author Contributions:** Conceptualization, Methodology, Writing paper, Funding Acquisition: S.R.; Software, Investigation, Data Curation: V.M. All authors have read and agreed to the published version of the manuscript.

**Funding:** This research is partially funded by the Ministry of Education, University, and Research (MIUR) of the Italian government.

**Conflicts of Interest:** The authors declare no conflict of interest.

## References


1. Dong, X.; Tan, C.; Dong, F. Gas-liquid two-phase flow velocity measurement with continuous wave ultrasonic doppler and conductance sensor. *IEEE Trans. Instrum. Meas.* **2017**, *66*, 3064–3076. [CrossRef]
2. Kotzé, R.; Ricci, S.; Birkhofer, B.; Wiklund, J. Performance tests of a new non-invasive sensor unit and ultrasound electronics. *Flow Meas. Instrum.* **2015**, *48*, 104–111. [CrossRef]
3. Evans, D.H.; McDicken, W.N. *Doppler Ultrasound Physics, Instrumentation and Signal Processing*; Wiley: Chichester, UK, 2000.
4. Wiklund, J.; Stading, M. Application of in-line ultrasound Doppler-based UVP–PD rheometry method to concentrated model and industrial suspensions. *Flow Meas. Instrum.* **2008**, *19*, 171–179. [CrossRef]
5. Newhouse, V.L.; Varner, L.W.; Bendick, P.J. Geometrical spectrum broadening in ultrasonic Doppler systems. *IEEE Trans. Biomed. Eng.* **1977**, *24*, 478–480. [CrossRef] [PubMed]
6. Newhouse, V.L.; Bendickand, P.J.; Varner, L.W. Analysis of transit time effects on Doppler flow measurement. *IEEE Trans. Biomed. Eng.* **1976**, *BME-23*, 381–387. [CrossRef] [PubMed]
7. Tortoli, P.; Guidi, G.; Newhouse, V.L. Improved blood velocity estimation using the maximum Doppler frequency. *Ultrasound Med. Biol.* **1995**, *21*, 527–532. [CrossRef]
8. Kathpalia, A.; Karabiyik, Y.; Eik-Nes, S.H.; Tegnander, E.; Ekroll, I.K.; Kiss, G.; Torp, H. Adaptive spectral envelope estimation for Doppler ultrasound. *IEEE Trans. Ultrason. Ferroelectr. Freq. Control* **2016**, *63*, 1825–1838. [CrossRef]
9. Ricci, S.; Vilkomerson, D.; Matera, R.; Tortoli, P. Accurate blood peak velocity estimation using spectral models and vector Doppler. *IEEE Trans. Ultrason. Ferroelectr. Freq. Control* **2015**, *62*, 686–696. [CrossRef]
10. Ricci, S.; Meacci, V.; Birkhofer, B.; Wiklund, J. FPGA-based System for In-Line Measurement of Velocity Profiles of Fluids in Industrial Pipe Flow. *IEEE Trans. Ind. Electron.* **2017**, *64*, 3997–4005. [CrossRef]
11. Jensen, J.A. *Estimation of Blood Velocities Using Ultrasound*; Cambridge University Press: Cambridge, UK, 1996.
12. Ricci, S. Switching power suppliers noise reduction in ultrasound Doppler fluid measurements. *Electronics* **2019**, *8*, 421. [CrossRef]
13. Ricci, S.; Bassi, L.; Boni, E.; Dallai, A.; Tortoli, P. Multichannel FPGA-based arbitrary waveform generator for medical ultrasound. *Electron. Lett.* **2007**, *43*, 1335–1336. [CrossRef]
14. Giannelli, P.; Bulletti, A.; Granato, M.; Frattini, G.; Calabrese, G.; Capineri, L. A Five-Level, 1–MHz, Class-D Ultrasonic driver for guided-wave transducer arrays. *IEEE Trans. Ultrason. Ferroelectr. Freq. Control* **2019**, *66*, 1616–1624. [CrossRef] [PubMed]
15. Gran, F.; Jakobsson, A.; Jensen, J.A. Adaptive spectral Doppler estimation. *IEEE Trans. Ultrason. Ferroelectr. Freq. Control* **2009**, *56*, 700–714. [CrossRef] [PubMed]
16. Ricci, S.; Meacci, V. Data-adaptive coherent demodulator for high dynamics pulse-wave ultrasound applications. *Electronics* **2018**, *7*, 434. [CrossRef]
17. Bjaerum, S.; Torp, H.; Kristoffersen, K. Clutter filter design for ultrasound color flow imaging. *IEEE Trans. Ultrason. Ferroelectr. Freq. Control* **2002**, *49*, 204–216. [CrossRef] [PubMed]
18. Tortoli, P.; Guidi, F.; Guidi, G.; Atzeni, C. Spectral velocity profiles for detailed ultrasound flow analysis. *IEEE Trans. Ultrason. Ferroelectr. Freq. Control.* **1996**, *43*, 654–659. [CrossRef]
19. Cooley, J.W.; Tukey, J.W. An algorithm for the machine calculation of complex Fourier series. *Math. Comput.* **1965**, *19*, 297–301. [CrossRef]
20. Ricci, S. Adaptive spectral estimators for fast flow profile detection. *IEEE Trans. Ultrason. Ferroelectr. Freq. Control.* **2013**, *60*, 421–427. [CrossRef]
21. Tronci, S.; Van Neer, P.; Giling, E.; Stelwagen, U.; Piras, D.; Mei, R.; Corominas, F.; Grosso, M. In-line monitoring and control of rheological properties through data-driven ultrasound soft-sensors. *Sensors* **2019**, *19*, 5009. [CrossRef]
22. Karabiyik, Y.; Ekroll, I.K.; Eik-Nes, S.H.; Avdal, J.; Løvstakken, L. Adaptive spectral estimation methods in color flow imaging. *IEEE Trans. Ultrason. Ferroelectr. Freq. Control.* **2016**, *63*, 1839–1851. [CrossRef]
23. Jensen, J.A.; Svendsen, N.B. Calculation of pressure fields from arbitrarily shaped, apodized, and excited ultrasound transducers. *IEEE Trans. Ultrason. Ferroelect. Freq. Control.* **1992**, *39*, 262–267. [CrossRef] [PubMed]
24. Jensen, J.A. Field: A program for simulating ultrasound systems. *Med. Biol. Eng. Comp.* **1996**, *34*, 351–353.

25. Wagner, R.F.; Smith, S.W.; Sandrik, J.M.; Lopez, H. Statistics of speckle in ultrasound B-scans. *IEEE Trans. Sonics Ultrason.* **1983**, *SU-30*, 156–163. [CrossRef]
26. Altera-Intel, FFT IP Core User Guide. Available online: https://www.intel.com/content/dam/www/programmable/us/en/pdfs/literature/ug/ug_fft.pdf (accessed on 7 March 2020).
27. Ramnarine, K.V.; Nassiri, D.K.; Hoskins, P.R.; Lubbers, J. Validation of a new blood-mimicking fluid for use in Doppler flow test objects. *Ultrasound Med. Biol.* **1998**, *24*, 451–459. [CrossRef]
28. Ricci, S.; Diciotti, S.; Francalanci, L.; Tortoli, P. Accuracy and reproducibility of a novel dual-beam vector doppler method. *Ultrasound Med. Biol.* **2009**, *35*, 829–838. [CrossRef]
29. Lui, E.Y.L.; Steinman, A.H.; Cobbold, R.S.C.; Johnston, K.W. Human factors as a source of error in peak Doppler velocity measurement. *J. Vasc. Surg.* **2005**, *42*, 972–979. [CrossRef]
30. Ricci, S.; Ramalli, A.; Bassi, L.; Boni, E.; Tortoli, P. Real-time blood velocity vector measurement over a 2-D. Region, *IEEE Trans. Ultrason. Ferroelect. Freq. Control.* **2018**, *65*, 201–209. [CrossRef]

*Article*

# An Interference Suppression Method for Non-Contact Bioelectric Acquisition

**Yue Tang, Ronghui Chang, Limin Zhang * and Feng Yan**

School of Electronic Science and Engineering, Nanjing University, Nanjing 210046, China;
dg1723024@smail.nju.edu.cn (Y.T.); mf1823002@smail.nju.edu.cn (R.C.); fyan@nju.edu.cn (F.Y.)
* Correspondence: lmzhang@nju.edu.cn

Received: 14 January 2020; Accepted: 6 February 2020; Published: 8 February 2020

**Abstract:** For non-contact bioelectrical acquisition, a new interference suppression method, named 'noise neutralization method', is proposed in this paper. Compared with the traditional capacitive driven-right-leg method, the proposed method is characterized with that there is an optimal gain to achieve the minimum interference output whatever for the electrode interface impedance mismatch caused by body motion and is more effective for smaller reference electrode areas. The performance of traditional capacitive driven-right-leg method is analyzed and the difficulty to suppress interference in the case of the interface impedance mismatch is pointed out. Therefore, a noise neutralization method is proposed by applying the reference electrode and a 50 Hz band-pass filter to obtain the interference of the human body and adapting the gains to neutralize the interference inputs of two acquisition electrodes and achieve the minimum interference output. The performance of the proposed method is theoretically analyzed and verified by the experiment results, which shows that the proposed method has similar performance to that of the traditional capacitive driven-right-leg method with electrode interface impedance match, while has better interference suppression ability with electrode interface impedance mismatch caused by body motion. It is suggested that the proposed method can be preferred in the case of limited reference electrode area or interface impedance mismatch.

**Keywords:** interference suppression; non-contact electrode; impedance mismatch; driven-right-leg; electrocardiogram; bioelectric acquisition

---

## 1. Introduction

It is well-known that, with the increasing demand for personal daily health monitoring, the study of wearable bioelectrical acquisition equipment is becoming increasingly popular. The quality of acquisition signals is the primary consideration, which is closely related to the condition of electrode interface. Needle electrodes or skin abrasions are required in early bioelectrical acquisition techniques in order to acquire high quality signals [1]. Nowadays, wet electrodes (i.e., Ag/AgCl electrodes) are usually used to acquire high quality signals. However, wet electrodes may cause skin irritation and allergic contact dermatitis [2–4], which leads to the difficulty for long-term bioelectricity acquisition. Therefore, capacitive electrodes are often used in wearable bioelectrical acquisition equipment [5–12]. Sun et al. defined the capacitive electrode as three types, dry electrodes, insulated electrodes, and non-contact electrodes [6]. Non-contact electrodes can measure surface potential through the clothes or other dielectrics, which can overcome the limitation of traditional wet electrode. To acquire high quality bioelectric signals by non-contact electrodes, various studies are carried out, involving electrode design and the front-end design [1–3,5,6,11]. It is pointed out that non-contact electrodes are very susceptible to power line common mode interference (CMI) due to the high impedance of electrodes [13–15], and larger CMI will cause lower signal to noise ratio. Moreover, the body motion may lead to interface impedance mismatch of the electrode, and part of CMI will be changed to pseudo difference mode components (PDMC) [6,16], this will degrade the signal quality further. Therefore, it is necessary to give more attention to suppress CMI of wearable bioelectric devices to acquire high quality signal [17–20].

For CMI suppression, the driven-right-leg (DRL) method has been widely used to suppress CMI by feeding back the reverse common mode signal from the front-end output to the driving electrode on the subject (human body) [8,19,21–23]. Our group Ding et al. proposed an improved front-end circuit with virtual DRL circuit and verified the design by three wearable ECG applications, portable finger ECG measurement, palm ECG acquisition for cycling and chest ECG test under different motion states [24]. Xu et al. proposed a front-end design to improve common mode rejection capability by combining both the feed-forward method and the DRL method [25]. Sakuma et al. proposed a circuit structure by connecting the driving electrode to the signal ground to suppress CMI [26]. For non-contact bioelectrical acquisition, Lim et al. proposed a capacitive ECG recording system with capacitive driven-right-leg (C-DRL) circuit integrated in the chair seat [15]. The C-DRL method is a variant of DRL method and is often used to improve common mode rejection ratio (CMRR) of non-contact bioelectrical acquisition, which is characterized with all electrodes contacting the body with isolation or clothes [14,19]. For C-DRL or DRL method, large feedback gain is usually required to obtain good suppression performance, this may destroy the stability of the front-end [21]. At the same time, large area of driving electrodes is often used in C-DRL or DRL method, which presents a difficulty of wearable device implementation [14,19,21].

On the other hand, PDMC is considered as part of motion artifacts. Serteyn et al. used an injection signal to track the change of coupling capacitance, so as to estimate and reduce motion artifacts in ECG measurement [27]. A three-axis accelerometer was used to record motion information for reducing motion artifacts reduction [28]. Rodrigues et al. uses neural networks to reconstruct ECG signals with severe motion artifacts [29]. The above methods are based on additional components or complex algorithms, which are difficult to implement.

For non-contact bioelectrical acquisition of wearable devices, a new interference rejection method, named noise neutralization method, is proposed in this paper. Compared with C-DRL or DRL method, the proposed method can effectively reduce the area of the driving electrode, and has a stronger ability to suppress interference CMI and PDMC with the electrode interface impedance mismatch caused by body motion. Firstly, the performance of traditional capacitive driven-right-leg method is analyzed and the difficulty to suppress interference in the case of the interface impedance mismatch is pointed out. Secondly, the noise neutralization method is proposed to suppress interference CMI and PDMC of non-contact electrodes and the performance is compared with that of C-DRL method. Finally, the feasibility of the proposed method is further verified by wearable ECG acquisition device.

## 2. Methods and Models

### 2.1. Traditional Capacitive Driven-Right-Leg Model for Eliminating Common Mode Interference

Figure 1 is an equivalent circuit model of the dual-electrode bioelectric acquisition equipment for CMI suppression with C-DRL method, where $V_P$ is the noise source of the power supply line, $V_{CM}$ is the CMI of the organism relative to the earth, $V'_{CM}$ is the CMI of the organism relative to signal ground, $C_P$ is the coupling capacitance between the power supply line and the organism, $C_B$ is the coupling capacitance between the earth and the organism, $C_S$ is the coupling capacitance between the earth and the signal ground, $Z_{E1}$ and $Z_{E2}$ are the interface impedances of the two acquisition electrodes, $Z_{E3}$ is the interface impedances of the driving electrode, $Z_{A1}$ and $Z_{A2}$ are the two equivalent input impedances of the front-ends, $V_{O1}$ and $V_{O2}$ are the outputs of the front-ends, and $k$ is the gain of the inversion amplifier.

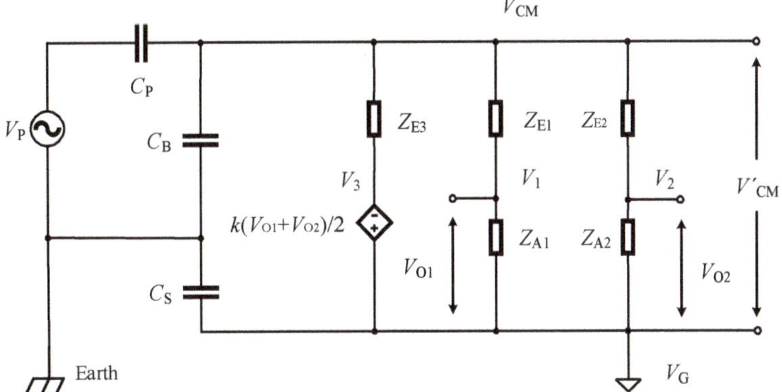

**Figure 1.** Equivalent circuit model of the dual-electrode bioelectric acquisition equipment for CMI suppression with C-DRL method.

According to Kirchhoff's current law, $V'_{CM}$, $V_{O1}$ and $V_{O2}$ can be expressed as

$$V'_{CM} = \frac{\frac{Z_B+Z_S}{Z_B}}{Z_P\left[\frac{1}{Z_P} + \frac{1}{Z_B} + \frac{1-A}{Z_{E1}} + \frac{1-B}{Z_{E2}} + \frac{1}{Z_{E3}}\left(1 - \frac{k}{2}(A+B) - (k-1)\frac{Z_S}{Z_B}\right)\right]}V_P, \tag{1}$$

$$V_{O1} = \frac{\frac{Z_{A1}(Z_B+Z_S)}{Z_B(Z_{E1}+Z_{A1})}}{Z_P\left[\frac{1}{Z_P} + \frac{1}{Z_B} + \frac{1-A}{Z_{E1}} + \frac{1-B}{Z_{E2}} + \frac{1}{Z_{E3}}\left(1 - \frac{k}{2}(A+B) - (k-1)\frac{Z_S}{Z_B}\right)\right]}V_P, \tag{2}$$

$$V_{O2} = \frac{\frac{Z_{A2}(Z_B+Z_S)}{Z_B(Z_{E2}+Z_{A2})}}{Z_P\left[\frac{1}{Z_P} + \frac{1}{Z_B} + \frac{1-A}{Z_{E1}} + \frac{1-B}{Z_{E2}} + \frac{1}{Z_{E3}}\left(1 - \frac{k}{2}(A+B) - (k-1)\frac{Z_S}{Z_B}\right)\right]}V_P, \tag{3}$$

$$A = \frac{Z_{A1}Z_B - Z_{E1}Z_S}{Z_{A1}Z_B + Z_{E1}Z_B} \ , \ B = \frac{Z_{A2}Z_B - Z_{E2}Z_S}{Z_{A2}Z_B + Z_{E2}Z_B} \ , \ k < 0 \tag{4}$$

when $Z_P$, $Z_B$, $Z_S$, $Z_{A1}$, $Z_{A2}$, $Z_{E1}$, and $Z_{E2}$ are invariant, it can be concluded from (1) that $V'_{CM}$ can be effectively decreased by reducing the interface impedance of the driving electrode ($Z_{E3}$) and increasing the gain of the inversion amplifier ($|k|$). However, too large a $|k|$ may lead to the output saturation of the inverted amplifier. When the value of $|k|$ is greater than 100, the performance will not be improved further [15]. Therefore, small $Z_{E3}$ is usually used to reduce $V'_{CM}$ in practical applications. There are several factors affecting the impedance of the coupling capacitance of the driving electrode. Generally, increasing the coupling area of the driving electrode is the easiest way to reduce $Z_{E3}$ and is widely used [2,15,16,19]. However, large area of the driving electrode is not convenient to minimize the wearable bioelectric acquisition equipment.

For the match of the electrode interface impedance, the output of differential amplifier $\Delta V_O$ can be written as

$$\Delta V_O = |V_{O1} - V_{O2}| = 0 \tag{5}$$

Considering the existence of interface impedance mismatch coefficient $\alpha$ between two electrodes, $Z_{E1}$ and $Z_{E2}$ can be expressed as

$$Z_E = Z_{E1} \ , \ Z_{E2} = (1+\alpha)Z_{E1}, \tag{6}$$

and for the mismatch of the electrode interface impedance, PDMC $\Delta V_O$ can be written as

$$\Delta V_O = \frac{\frac{Z_B+Z_S}{Z_B}\left|\frac{Z_A}{Z_A+Z_E} - \frac{Z_A}{Z_A+(1+\alpha)Z_E}\right|}{Z_P\left[\frac{1}{Z_P} + \frac{1}{Z_B} + \frac{1-A}{Z_E} + \frac{1-B}{(1+\alpha)Z_E} + \frac{1}{Z_{E3}}\left(1 - \frac{k}{2}(A+B) - (k-1)\frac{Z_S}{Z_B}\right)\right]}V_P \tag{7}$$

If the area ratio of the acquisition electrode to the driving electrode is $\beta$, the interface impedance of the driving electrode satisfies

$$Z_{E3} = \beta Z_E, \tag{8}$$

The analysis can be further simplified by

$$Z_A = Z_{A1} = Z_{A2}, \tag{9}$$

$$Z_E = Z_{E1} = Z_{E2}, \tag{10}$$

For the non-contact electrode based on cotton material, the electrode interface impedance can be assumed as (11) [10].

$$Z_E = 305 \text{ M}\Omega \| 34 \text{ pF} \tag{11}$$

Considering the power line with the amplitude 220 V and the frequency 50 Hz, the following relationship is reasonable [16,22].

$$Z_P = 100Z_B = 100Z_S = 1.6 \text{ G}\Omega \tag{12}$$

Generally, different front-end equivalent input impedances are required for different electrode interfaces in order to acquire high quality signals. According to American Standards [30], the relationship between the electrode interface impedance, the equivalent input impedance and the coupled capacitor impedance can be expressed by (13) and (14),

$$Z_A = 6.7Z_E, \tag{13}$$

$$Z_B = 0.05Z_E. \tag{14}$$

and PDMC $\Delta V_O$ of non-contact electrodes can be derived from (9)–(14).

$$\Delta V_O = \frac{\left|\frac{6.7\alpha}{7.7(7.7+\alpha)}\right|}{\left[50.5 + \frac{5}{7.7} + \frac{5}{7.7+\alpha} + \frac{5}{\beta}\left(1 - \frac{k}{2}\left(\frac{13.4}{7.7+\alpha} + \frac{6.7\alpha}{7.7(7.7+\alpha)}\right)\right)\right]}V_P \tag{15}$$

Figure 2 shows PDMC $\Delta V_O$ as function of the impedance mismatch coefficient of two acquisition electrodes with different areas of the driving electrode for C-DRL method, where the value of $|k|$ is 100 according to the setting in reference [16]. It can be seen that $\Delta V_O$ increases with the increase of impedance mismatch coefficient $\alpha$ for constant $\beta$, decreases with the decrease of $\beta$ for constant $\alpha$, which means large area of the driving electrode is expected and this will limit the miniaturization of the wearable acquisition equipment.

Figure 3 shows PDMC $\Delta V_O$ as functions of the gain of the inversion amplifier and the impedance mismatch coefficient of two acquisition electrodes for C-DRL method, where the blue surface represents the case of $\beta$ being 1/30, the red surface represents the case of $\beta$ being 1 and the yellow surface represents the case of $\beta$ being 20. It can be seen that $\Delta V_O$ increases with the increase of the impedance mismatch coefficient $\alpha$ for constant $\beta$ and $|k|$, decreases with the increase of $|k|$ in a certain range and is almost unchanged for $|k|$ larger than a certain value. $\Delta V_O$ can be further suppressed by the decrease of $\beta$ for constant $|k|$, which agrees on the performance shown in Figure 3. Therefore, it is difficult for C-DRL method to suppress CMI with small area of the driving electrode, especially under the situation of the impedance mismatch of two acquisition electrodes. Therefore, a noise neutralization method is proposed to solve above difficulties.

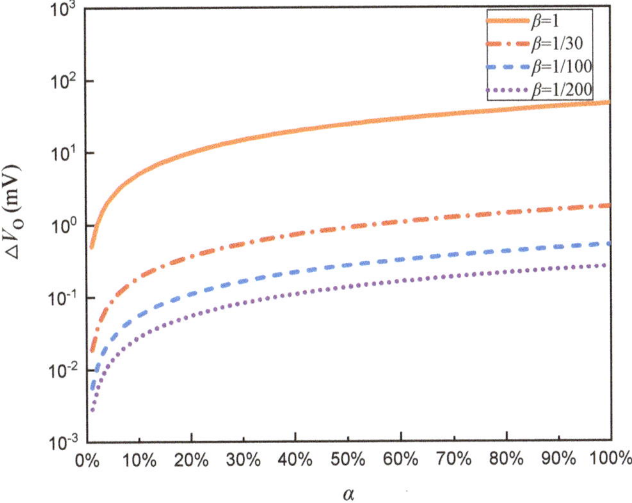

**Figure 2.** PDMC $\Delta V_O$ as function of the impedance mismatch coefficient of two acquisition electrodes with different areas of the driving electrode for C-DRL method, where the value of $|k|$ is 100 according to the setting in [16]. It increases with the increase of the impedance mismatch coefficient $\alpha$ for constant $\beta$ and $|k|$. PDMC can be further suppressed by the decrease of $\beta$ for constant $|k|$.

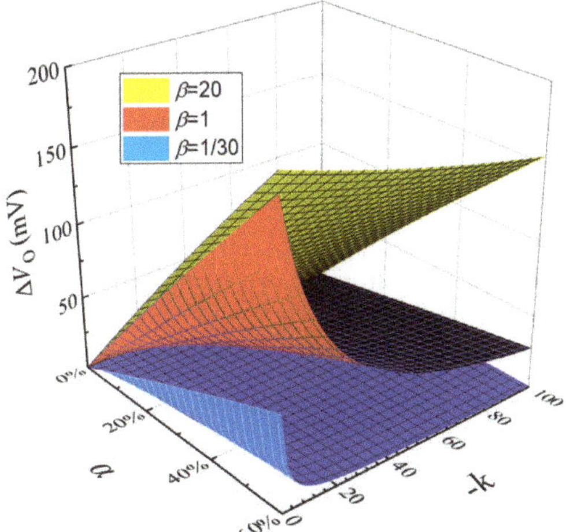

**Figure 3.** PDMC $\Delta V_O$ as functions of the gain of the inversion amplifier and the impedance mismatch coefficient of two acquisition electrodes for C-DRL method. It increases with the increase of the impedance mismatch coefficient $\alpha$ for constant $\beta$ and $|k|$, decreases with the increase of $|k|$ in a certain range and is almost unchanged for $|k|$ larger than a certain value. PDMC can also be further suppressed by the decrease of $\beta$ for constant $|k|$.

*2.2. Noise Neutralization Method to Eliminate Common Mode Interference Model*

Schematic diagram of the dual-electrode bioelectric acquisition equipment for CMI suppression with the proposed noise neutralization method is shown in Figure 4, including two acquisition electrodes ($Z_{E1}$ and $Z_{E2}$ are the interface impedances) and corresponding front-ends $A_1$ and $A_2$ ($Z_{A1}$ and $Z_{A2}$ are the equivalent input impedances), a reference electrode ($Z_{E3}$ is the interface impedance), and corresponding front-end $A_3$ ($Z_{A3}$ is the equivalent input impedance) and processing circuits. The processing circuits include a 50 Hz band-pass filter, two variable gain amplifiers ($k_1$ and $k_2$ are the gains), a micro control unit (MCU), an analog to digital converter (ADC), and a differential amplifier. In addition, the parameters explanations ($V_P$, $V_{CM}$, $V'_{CM}$, $C_P$, $C_B$, and $C_S$) are the same with that in Figure 1. The dotted line area is the neutral part, which uses the reference electrode and a 50 Hz band-pass filter to obtain the CMI of the human body and adapts the gains $k_1$ and $k_2$ to vary the CMI amplitude to the input ends of the differential amplifier in order to neutralize the CMI from two acquisition electrodes and achieve the minimum CMI output of the differential amplifier. It should be noted that the signal from the acquisition electrode will be attenuated by two cascading resistors $Z_M$ and $Z_N$. Small attenuation requires large ratio of $Z_N$ to $Z_M$. In the following analysis, the ratio of $Z_N$ to $Z_M$ is set as 9.

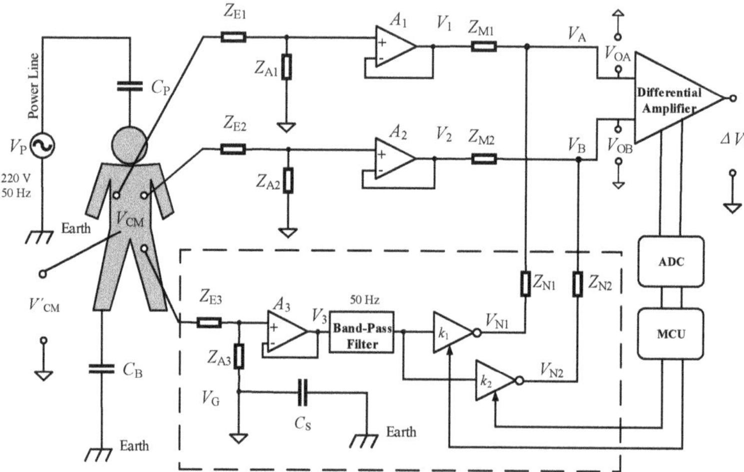

**Figure 4.** Schematic diagram of the dual-electrode bioelectric acquisition equipment for CMI suppression with the proposed noise neutralization method.

Figure 5 is the equivalent circuit model of the proposed neutralization method.

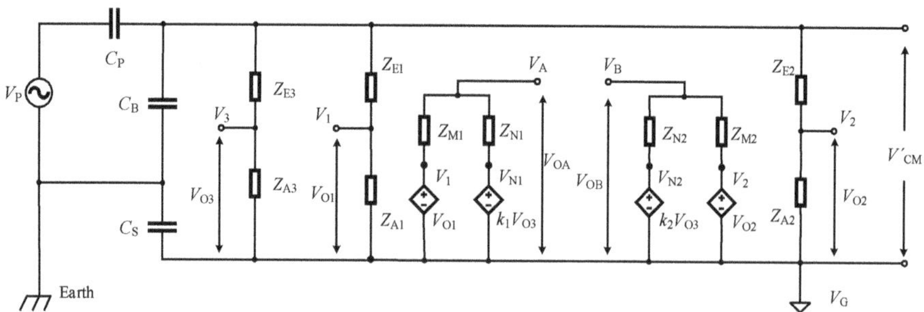

**Figure 5.** Equivalent circuit model of the dual-electrode bioelectric acquisition equipment for CMI suppression with the proposed noise neutralization method.

According to Kirchhoff's current law, the CMI of the organism relative to signal ground $V'_{\text{CM}}$, and the outputs of the front-ends $V_{\text{OA}}$ and $V_{\text{OB}}$ can be expressed as

$$V'_{\text{CM}} = \frac{\frac{Z_B + Z_S}{Z_B}}{Z_P\left[\frac{1}{Z_P} + \frac{1}{Z_B} + \frac{1-A}{Z_{E1}} + \frac{1-B}{Z_{E2}} + \frac{1-C}{Z_{E3}}\right]} V_P, \tag{16}$$

$$V_{\text{OA}} = \frac{\frac{Z_{N1}}{Z_{M1}+Z_{N1}}\left(A + \frac{Z_S}{Z_B}\right) + \frac{Z_{M1}}{Z_{M1}+Z_{N1}}k_1\left(C + \frac{Z_S}{Z_B}\right)}{Z_P\left[\frac{1}{Z_P} + \frac{1}{Z_B} + \frac{1-A}{Z_{E1}} + \frac{1-B}{Z_{E2}} + \frac{1-C}{Z_{E3}}\right]} V_P, \tag{17}$$

$$V_{\text{OB}} = \frac{\frac{Z_{N2}}{Z_{M2}+Z_{N2}}\left(B + \frac{Z_S}{Z_B}\right) + \frac{Z_{M2}}{Z_{M2}+Z_{N2}}k_2\left(C + \frac{Z_S}{Z_B}\right)}{Z_P\left[\frac{1}{Z_P} + \frac{1}{Z_B} + \frac{1-A}{Z_{E1}} + \frac{1-B}{Z_{E2}} + \frac{1-C}{Z_{E3}}\right]} V_P, \tag{18}$$

$$A = \frac{Z_{A1}Z_B - Z_{E1}Z_S}{Z_{A1}Z_B + Z_{E1}Z_B}, B = \frac{Z_{A2}Z_B - Z_{E2}Z_S}{Z_{A2}Z_B + Z_{E2}Z_B}, C = \frac{Z_{A3}Z_B - Z_{E3}Z_S}{Z_{A3}Z_B + Z_{E3}Z_B} \tag{19}$$

As can be seen from (17)–(19), the common component of $V_{\text{OA}}$ and $V_{\text{OB}}$ can be reduced to 0 by adjusting the gains $k_1$ and $k_2$ to optimal values, and the CMI will be suppressed fully. The optimal $k_1$ and $k_2$ can be expressed as $k_{\text{optimal-1}}$ and $k_{\text{optimal-2}}$ by (20) and (21).

$$k_{\text{optimal-1}} = -\frac{Z_{N1}}{Z_{M1}}\frac{Z_{A1}(Z_{A3} + Z_{E3})}{Z_{A3}(Z_{A1} + Z_{E1})}, \tag{20}$$

$$k_{\text{optimal-2}} = -\frac{Z_{N2}}{Z_{M2}}\frac{Z_{A2}(Z_{A3} + Z_{E3})}{Z_{A3}(Z_{A2} + Z_{E2})}, \tag{21}$$

Under the situation that the neutralizing resistances $Z_{N1}$, $Z_{N2}$, $Z_{M1}$, $Z_{M2}$ and the equivalent input impedances $Z_{A1}$, $Z_{A2}$, $Z_{A3}$ are definite values, $k_{\text{optimal-1}}$ and $k_{\text{optimal-2}}$ are only related to $Z_{E1}$, $Z_{E2}$, and $Z_{E3}$ according to (20) and (21). The PDMC output of the differential amplifier can be expressed as (22), which shows that $\Delta V_O$ is related to the impedance mismatch coefficient $\alpha$ of the electrode interface, the area ratio $\beta$ of the acquisition electrode to the reference electrode, and the gain difference $|\Delta k|$. It should be noted that there is an optimal $\Delta k$, defined $\Delta k_{\text{optimal}}$, can still make $\Delta V_O$ be equal to 0 even if $k_1$ and $k_2$ cannot satisfy (20) and (21).

$$\Delta V_O = \left|\frac{\frac{Z_{N1}}{Z_{M1}+Z_{N1}}(A - B) + \frac{Z_{M1}}{Z_{M1}+Z_{N1}}\Delta k\left(C + \frac{Z_S}{Z_B}\right)}{Z_P\left[\frac{1}{Z_P} + \frac{1}{Z_B} + \frac{1-A}{Z_{E1}} + \frac{1-B}{Z_{E2}} + \frac{1-C}{Z_{E3}}\right]}\right| V_P, \Delta k = k_1 - k_2 \tag{22}$$

The performance comparison between C-DRL method and proposed noise neutralization method is shown in Figure 6, where the blue surface represents C-DRL method with $\beta$ being 1 (the area of the acquisition electrode is the same to that of the driving electrode), and the red surface represents the proposed method with $\beta$ being 1, the violet surface represents C-DRL method with $\beta$ being 1/30 (the area of driving electrode is 30 times of the area of acquisition electrode) and the yellow surface represents the proposed method with $\beta$ being 20 (the area of acquisition electrode is 20 times of the area of reference electrode). Considering $k$ less than 0, $|k|$ and $|\Delta k|$ have been replaced by $-k$ and $\Delta k$ respectively to unify the coordinate axes of the chart.

When the impedance mismatch coefficient $\alpha$ exists, $\Delta k$ in a certain range of the proposed method can obtain a smaller PDMC than that of C-DRL method. Even at the case of $\beta$ being 1/30 for C-DRL method and $\beta$ being 1 for proposed method, this conclusion can still be reached. Moreover, for the proposed method, the range of $\Delta k$ can be larger for $\beta$ being 20 than that for $\beta$ being 1, which means the proposed method is more effective than that of the C-DRL method, especially for small area of the reference electrode.

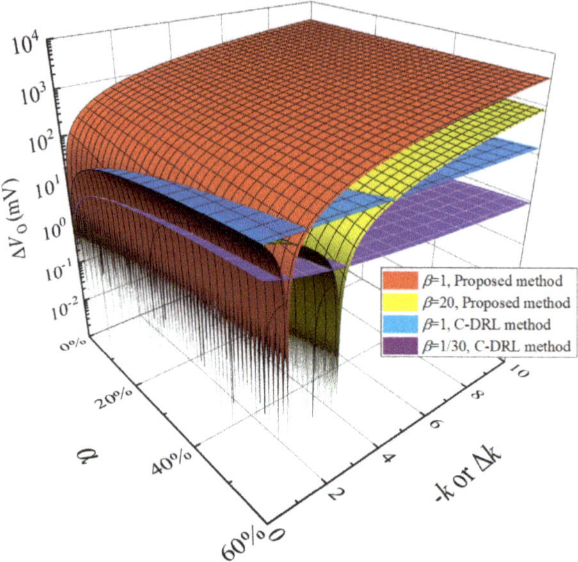



**Figure 6.** The performance comparison between C-DRL method and the proposed noise neutralization method, where the red and yellow surfaces represent the proposed method with $\beta$ being 1 and 20, respectively. The blue and violet surfaces represent the C-DRL method with $\beta$ being 1 and 1/30, respectively. When the impedance mismatch coefficient $\alpha$ exists, $\Delta k$ in a certain range of the proposed method can obtain a smaller PDMC than that of C-DRL method and there always exists a certain smaller optimal gain to minimize PDMC. Theoretically, under the same conditions, the proposed method has better PDMC suppression with a larger $\beta$ (smaller reference electrode area) and a smaller gain (compared with the typical driving gain of C-DRL method).

## 3. Experiments

The circuit system for the proposed neutralization method is implemented with two-board components, as shown in Figure 7, where the left side is the acquisition motherboard and the right side is the noise neutralization board. In the acquisition motherboard, the front-end is designed according to the proposed structure in reference [31], the integrated analog to digital converter ADS1298 with 24-bit resolution is selected as the differential amplifier, and the microcontroller MSP430F5528 is used to sample the data from ADS1298 and transform the data thorough the series port to the application in the computer. In the noise neutralization board, the ratio of $Z_N$ to $Z_M$ is set as 2 ($Z_N$ being10 k$\Omega$ and $Z_M$ being 5.1 k$\Omega$) for easy resistance selection and the programmable resistor MAX5496 (10 k$\Omega$, 1024 taps) is used to obtain high precision gain to meet the requirement of $k_{optimal}$. It should be noted that the method for searching the optimal value adopted in this paper is dichotomy, which has three main steps. In the first step, the initial values of $k_1$ and $k_2$ are set respectively by controlling the taps of variable programmable resistors and the initial amplitudes of $V_{OA}$ and $V_{OB}$ are measured by the channel 2 and channel 3 of ADS1298, as shown in Figure 7. In the second step, the values of $k_1$ and $k_2$ are set to the median of the initial ranges and the amplitudes of $V_{OA}$ and $V_{OB}$ are measured again. In the third step, the values of $k_1$ and $k_2$ are reset according to the amplitudes of $V_{OA}$ and $V_{OB}$ measured in the previous two steps. These three steps will be repeated until the amplitudes of $V_{OA}$ and $V_{OB}$ are minimized. Applying the method, $k_{optimal}$ can be approximated and the amplitude of $\Delta V_O$ can be reduced effectively.

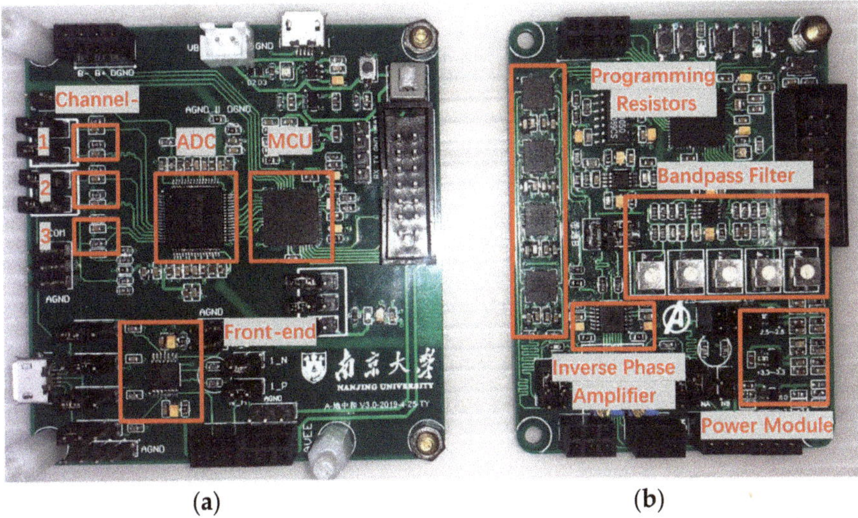



(a)        (b)

**Figure 7.** Circuit system for the proposed neutralization method, (**a**) the acquisition motherboard, (**b**) the noise neutralization board.

Figure 8 is the comparison of the frequency response characteristics between the acquisition system with C-DRL method and the acquisition system with the proposed method. They have similar frequency response characteristics for using the same configuration of the acquisition front-end.

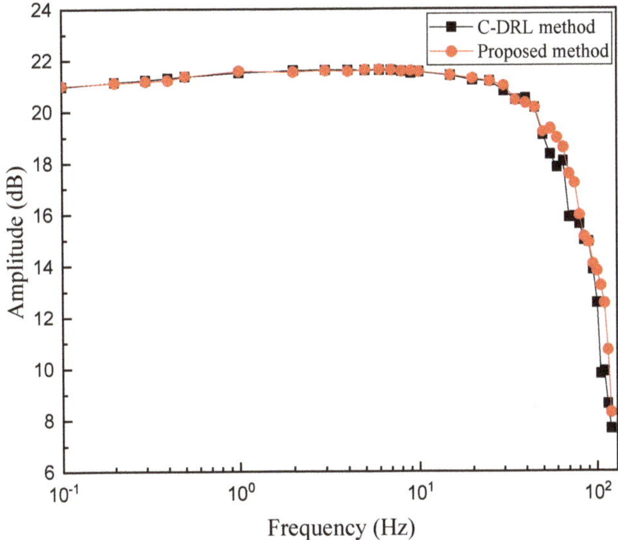



**Figure 8.** Comparison of the frequency response between the acquisition system with C-DRL method and the acquisition system with proposed method.

CMRR performance comparison between C-DRL method and the proposed method under different simulated impedance mismatch is shown in Table 1, where two input signals with same phase and different amplitude are used to simulate the electrode interface impedance mismatch. For $\alpha$ being 0, CMRRs for two methods are both about 90 dB. For $\alpha$ being 20%, CMRR for the proposed method can also achieve 90 dB by adjusting the gains $k_1$ and $k_2$ to optimal values, which means that the proposed method can suppress PDMC caused by the electrode mismatch effectively. However, for $\alpha$ being 20%, CMRR for C-DRL method is only 14.4 dB with the lack of feedback and may be improved with the closed-loop feedback. Therefore, the experiments with C-DRL method and the proposed neutralization method for human body ECG acquisition are carried out for further performance comparison.

**Table 1.** CMRR comparison of circuits using proposed method and C-DRL method.

| Simulation of Electrode Mismatch | Input Signal Amplitude of Electrode [1] | CMRR | |
|---|---|---|---|
| | | C-DRL Method | Proposed Method |
| $\alpha = 0$ | P: 500 mV<br>N: 500 mV | 91.45 dB | 90.36 dB |
| $\alpha = 20\%$ | P: 500 mV<br>N: 400 mV | 14.40 dB [2] | 90.22 dB |

[1] The frequency of all input signal is 50 Hz. For the proposed method, the input signal amplitude of the reference electrode R is always 500 mV. [2] Due to the lack of negative feedback, this simulation value is lower than the actual value of C-DRL method.

Figure 9 shows the devices and bandage electrodes for simultaneously human ECG acquisition with the C-DRL method and the proposed method. The blue marks the device and the electrodes with C-DRL method, and the red marks the device and the electrodes with the proposed method. The driving electrode D of C-DRL method and the reference electrode R of the proposed method have the same area and the dimensions are $3 \times 4$ cm. The areas of the acquisition electrodes P and N are designed according to the experiment requirements.

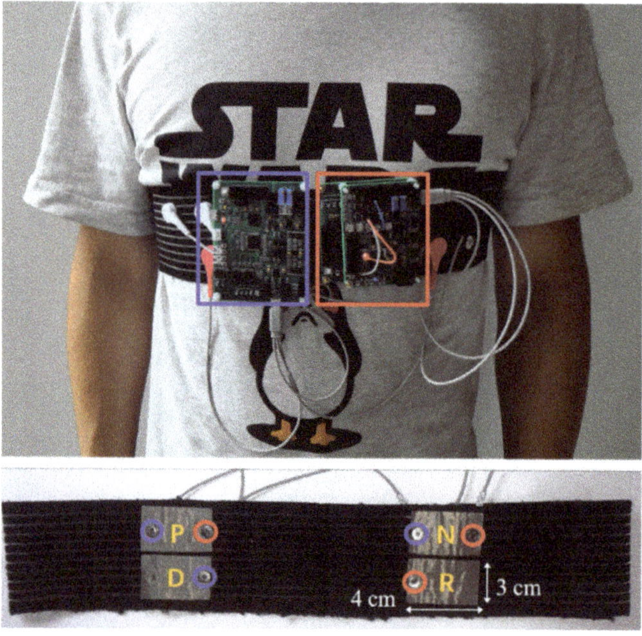


**Figure 9.** Devices and bandage electrodes for human body ECG acquisition using the C-DRL method and the proposed method.

When two acquisition devices are used to acquire ECG signal from human body at the same time, CMI coupled on human body may be reduced by C-DRL method, thus affecting the performance of the proposed method. Therefore, two devices are used to acquire ECG signal at the same chest position independently. For C-DRL method, the value of $k$ is −100. For the proposed method, the band-pass filter is designed with the center frequency being 50 Hz, the gain at the center frequency being 0 dB, and the quality factor being 5. The human test subject is the volunteer from our group, one male student aged 25 without known cardiac pathology.

Under static condition, the independent measurement results of human chest ECG acquisition with interface impedance match ($\alpha = 0$, the areas of electrodes P and N are both 3 × 4 cm) are shown in Figure 10 for two methods, where both the original signal and the ECG signal processed with 50 Hz digital notch filter from the original signal are given, showing that both two methods can obtain high quality ECG signals independently.

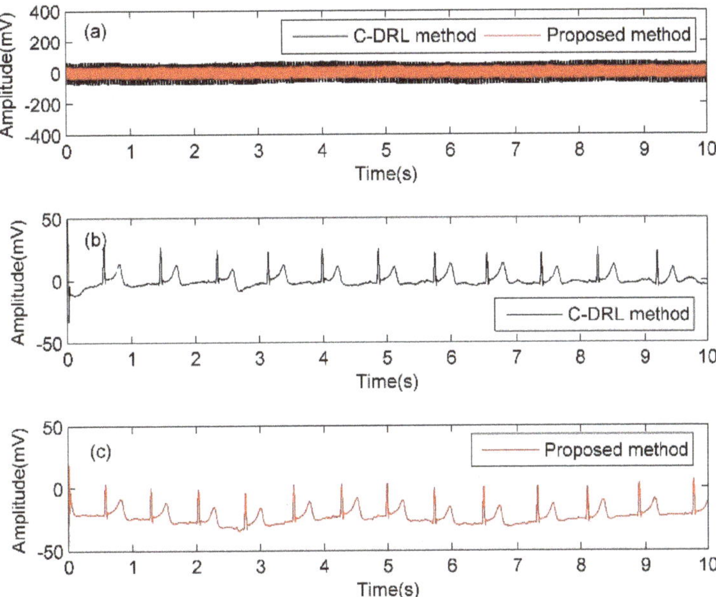

**Figure 10.** Independent measurement results of human chest ECG acquisition with interface impedance match under static condition, (**a**) time-domain waveform of original signal, (**b**) ECG signal of C-DRL method processed by software notch, (**c**) ECG signal of proposed method processed by software notch.

Subsequently, the independent measurement results of human chest ECG acquisition with interface impedance mismatch are carried out by using the insulated plastic to stem and reduce the area of electrode N artificially and simulate the impedance mismatch ($\alpha = 20\%$, the area of electrode P is 3 × 4 cm and the area of electrode N is about 3 × 3.2 cm), as shown in Figure 11, where the original signal amplitude of the C-DRL method is about 406 mV, while that of the proposed method is about 276 mV.

According to the results of Figures 10 and 11, the proposed method has similar performance to that of C-DRL method with electrode interface impedance match, while has better PDMC suppression ability than that of C-DRL method with electrode interface impedance mismatch.

Furthermore, in order to compare the performance of two methods at the same time, two devices are used to acquire ECG signal at the same time as shown in Figure 9. Similar to the steps of independent measurement mentioned above, under static condition, the simultaneous measurement results of human chest ECG acquisition with interface impedance match ($\alpha = 0$, the areas of electrode P and electrode N is are both 3 × 4 cm) are shown in Figure 12 and the simultaneous measurement results of human chest ECG acquisition with interface impedance mismatch ($\alpha = 20\%$, simulated, the area of electrode P is 3 × 4 cm and the area of electrode N is about 3 × 3.2 cm) are shown in Figure 13.

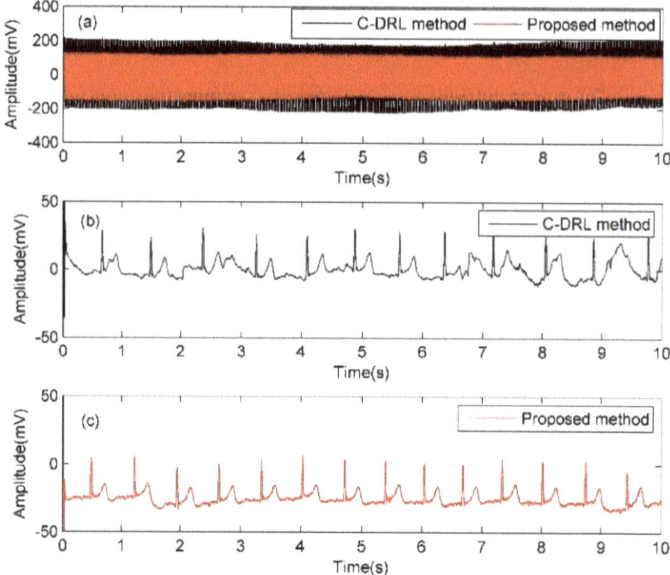

**Figure 11.** Independent measurement results of human chest ECG acquisition with interface impedance mismatch under static condition, (**a**) time-domain waveform of original signal, (**b**) ECG signal of C-DRL method processed by software notch, (**c**) ECG signal of the proposed method processed by software notch.

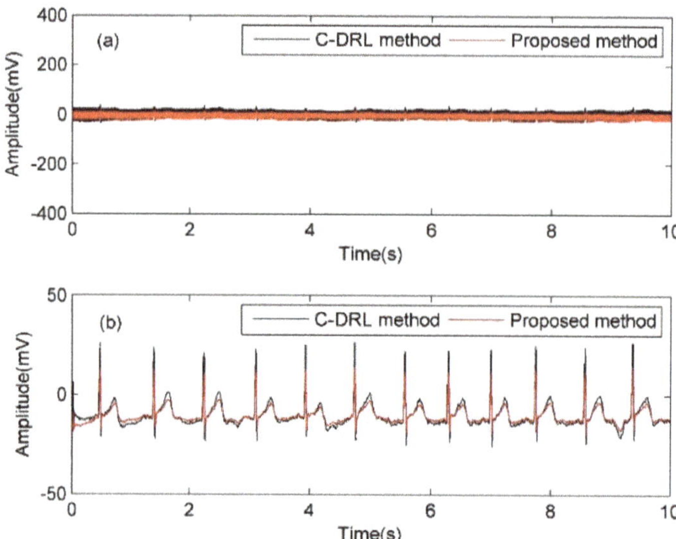

**Figure 12.** Simultaneous measurement results of human chest ECG acquisition with interface impedance match under static condition, (**a**) time-domain waveform of original signal, (**b**) ECG signal processed by software notch.

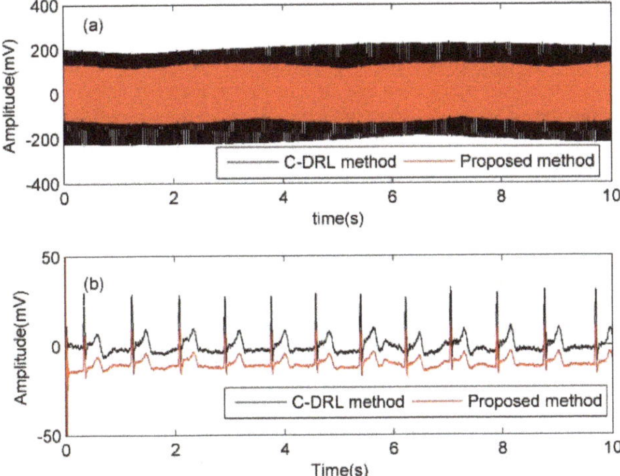

**Figure 13.** Simultaneous measurement results of human chest ECG acquisition with interface impedance mismatch under static condition, (**a**) time-domain waveform of original signal, (**b**) ECG signal processed by software notch.

According to the results of Figure 12, it can be concluded that the proposed method has similar performance to that of the C-DRL method with electrode interface impedance match. As shown in Figure 13, the original signal amplitude of the C-DRL method is about 405 mV, while the original signal amplitude of the proposed method is about 249 mV, which shows the proposed method has better PDMC suppression performance in the case of electrode interface impedance mismatch.

When the areas of electrodes P and N are the same, ECG measurements of the two devices at the same time with small amplitude swing of upper limb are shown in Figure 14, where the performances of two methods are alike for similar amplitudes of two original signals.

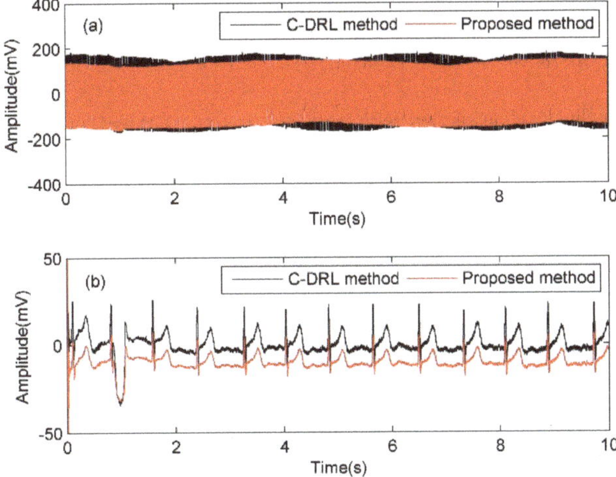

**Figure 14.** Simultaneous measurement results of human chest ECG acquisition with small amplitude swing of upper limb, (**a**) time-domain waveform of original signal, (**b**) ECG signal processed by software notch.

In order to clearly observe the electrode interface impedance mismatch caused by body motion, the upper limb stays still first, then rotates to one side substantially, and then stays still again. The simultaneous measurement results are shown in Figure 15, where two methods can both acquire clear ECG signals for the upper limb being still and the performances of two methods are degraded by the upper limb rotation. This can be explained in that the upper limb rotation leads to the electrode interface impedance mismatch, which causes the increase of PDMC. For C-DRL method, the amplitude of the original signal is about 157.5 mV before the upper limb rotation, and is about 838 mV after the upper limb rotation. For the proposed method, the amplitude of the original signal is about 190.4 mV before the upper limb rotation, and is about 530.1 mV after the upper limb rotation. The upper limb rotation causes 12.7 dB interference increase for C-DRL method, and about 5.0 dB interference increase for the proposed method, which also shows that the proposed method has better PDMC suppression ability in the case of the electrode interface impedance mismatch.

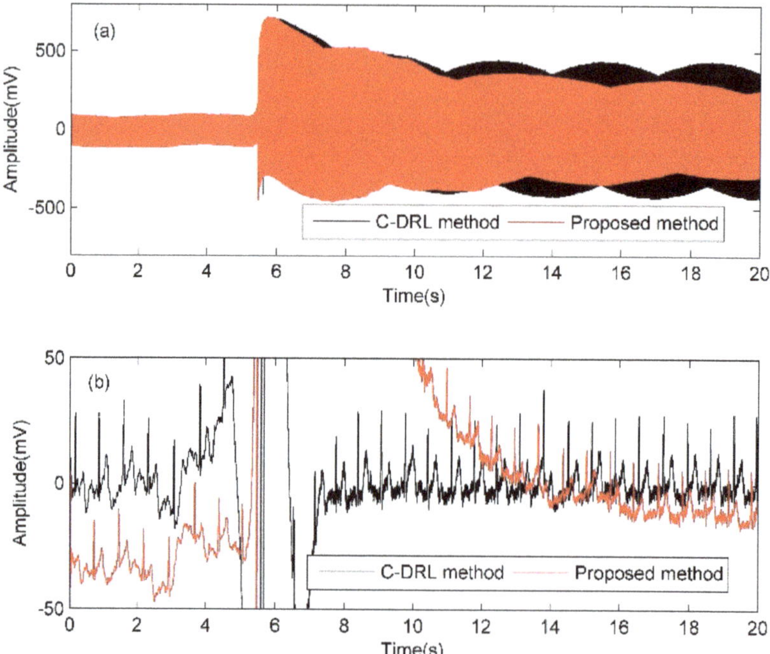


**Figure 15.** Simultaneous measurement results of human chest ECG acquisition for the process of the upper being still, rotating to one side substantially and being still again, (**a**) time-domain waveform of original signal, (**b**) ECG signal processed by software notch.

## 4. Discussion and Conclusions

For non-contact bioelectrical acquisition, a new interference suppression method, named 'noise neutralization method', is proposed in this paper. Compared with C-DRL or DRL method, the proposed method can effectively reduce the area of the driving electrode, and has a stronger ability to suppress interference, especially for the situation of electrode interface impedance mismatch caused by body motion.

Firstly, the performance of C-DRL method for non-contact acquisition is analyzed using the equivalent circuit model, which shows that interference suppression degrades with the decrease of the gain of the inversion amplifier and the area of driving electrode in the case of the interface impedance mismatch of the acquisition electrode. Therefore, a noise neutralization method is proposed to suppress

CMI and PDMC for non-contact bioelectrical acquisition by applying the reference electrode and a 50 Hz band-pass filter to obtain the CMI of the human body and adapting the gain $k_1$ and $k_2$ to neutralize the CMI input of two acquisition electrodes and achieve the minimum CMI output. The proposed method is characterized with that there is an optimal gain to achieve the minimum interference output whatever for the electrode interface impedance mismatch and is more effective for the smaller reference electrode area.

Subsequently, non-contact ECG acquisition devices based on C-DRL method and proposed method are designed to verify the performance of the proposed method. From the experiment results, it can be concluded that the proposed method has similar performance to that of the C-DRL method with electrode interface impedance match, while it has better PDMC suppression ability using a smaller reference electrode area and a smaller gain. This is the outstanding characterization of the proposed method. In general, the proposed method is an effective method to suppress interference CMI and PDMC under the situation of limited electrode area and unavoidable body motion.

It should be noted that the optimization method for adapting the gains $k_1$ and $k_2$ to optimal values is dichotomous in this paper, it is an effective method but its adapting time is not enough fast to suppress PDMC caused by strong body motion. Moreover, the variable gain amplifier with high programmable precision is required to obtain the optimal $k$, which may create complexity of the circuit structure.

In the future, a more effective adapting method will be studied to reduce the adapting time and the circuit structure will be improved for wearable applications.

**Author Contributions:** Conceptualization, L.Z.; Data curation, L.Z.; Funding acquisition, L.Z. and F.Y.; Investigation, Y.T. and R.C.; Project administration, L.Z. and F.Y.; Software, Y.T. and R.C.; Writing—original draft, Y.T. and R.C.; Writing—review and editing, L.Z. and F.Y. All authors have read and agreed to the published version of the manuscript.

**Funding:** This research was supported by National Key R&D Program of China (no. 2016YFA0202102) and National Nature Science Foundation Program of China (no. 11304152 and 61571376).



## Abbreviations

| | |
|---|---|
| ADC | Analog to digital converter |
| Ag/AgCl | Silver/silver chloride |
| C-DRL | Capacitive driven-right-leg |
| CMI | Common mode interference |
| CMRR | Common mode rejection ratio |
| DRL | Driven-right-leg |
| ECG | Electrocardiography |
| MCU | Micro control unit |
| PDMC | Pseudo difference mode components |

## References

1. Spinelli, E.; Haberman, M. Insulating electrodes: A review on biopotential front ends for dielectric skin-electrode interfaces. *Physiol. Meas.* **2010**, *31*, 183–198. [CrossRef] [PubMed]
2. Leicht, L.; Eilebrecht, B.; Weyer, S.; Leonhardt, S.; Teichmann, D. Closed-loop control of humidification for artifact reduction in capacitive ECG measurements. *IEEE Trans. Biomed. Circuits Syst.* **2017**, *11*, 300–313. [CrossRef]
3. Meziane, N.; Webster, J.G.; Attari, M.; Nimunkar, A.J. Dry electrodes for electrocardiography. *Physiol. Meas.* **2013**, *34*, 47–69. [CrossRef] [PubMed]
4. Boehm, A.; Yu, X.; Neu, W.; Leonhardt, S.; Teichmann, D. A Novel 12-Lead ECG T-Shirt with Active Electrodes. *Electronics* **2016**, *5*, 75. [CrossRef]
5. Acar, G.; Ozturk, O.; Golparvar, A.J.; Elboshra, T.A.; Böhringer, K.; Yapici, M.K. Wearable and Flexible Textile Electrodes for Biopotential Signal Monitoring: A review. *Electronics* **2019**, *8*, 479. [CrossRef]

6.  Sun, Y.; Yu, X.B. Capacitive biopotential measurement for electrophysiological signal acquisition: A review. *IEEE Sens. J.* **2016**, *16*, 2832–2853. [CrossRef]

7.  Oehler, M.; Ling, V.; Melhorn, K.; Schilling, M. A multichannel portable ECG system with capacitive sensors. *Physiol. Meas.* **2008**, *29*, 783–793. [CrossRef]

8.  Wannenburg, J.; Malekian, R.; Hancke, G.P. Wireless capacitive based ECG sensing for feature extraction and mobile health monitoring. *IEEE Sens. J.* **2018**, *18*, 6023–6032. [CrossRef]

9.  Yang, B.; Dong, Y.; Hou, Z.; Xue, X. Simultaneously capturing electrocardiography and impedance plethysmogram signals from human feet by capacitive coupled electrode system. *IEEE Sens. J.* **2017**, *17*, 5654–5662. [CrossRef]

10. Chi, Y.M.; Jung, T.; Cauwenberghs, G. Dry-contact and noncontact biopotential electrodes: Methodological review. *IEEE Rev. Biomed. Eng.* **2010**, *3*, 106–119. [CrossRef]

11. Tomasini, M.; Benatti, S.; Milosevic, B.; Farella, E.; Benini, L. Power line interference removal for high-quality continuous biosignal monitoring with low-power wearable devices. *IEEE Sens. J.* **2016**, *16*, 3887–3895. [CrossRef]

12. Sullivan, T.J.; Deiss, S.R.; Jung, T.P.; Cauwenberghs, G. A brain-machine interface using dry-contact, low-noise EEG sensors. In Proceedings of the IEEE International Symposium on Circuits and Systems (ISCAS 2008), Seattle, WA, USA, 18–21 May 2008; pp. 1986–1989.

13. Kim, K.K.; Park, K.S. Effective coupling impedance for power line interference in capacitive-coupled ECG measurement system. In Proceedings of the International Conference on Information Technology and Applications in Biomedicine (ITAB 2008), Shenzhen, China, 30–31 May 2008; pp. 256–258.

14. Lim, Y.G.; Kim, K.K.; Park, K.S. ECG measurement on a chair without conductive contact. *IEEE Trans. Biomed. Eng.* **2006**, *53*, 956–959. [PubMed]

15. Lim, Y.G.; Chung, G.S.; Park, K.S. Capacitive driven-right-leg grounding in indirect-contact ECG measurement. In Proceedings of the Annual International Conference of the IEEE Engineering in Medicine and Biology (EMBS 2010), Buenos Aires, Argentina, 31 August–4 September 2010; pp. 1250–1253.

16. Kim, K.K.; Lim, Y.G.; Park, K.S. Common mode noise cancellation for electrically non-contact ECG measurement system on a chair. In Proceedings of the IEEE Engineering in Medicine and Biology 27th Annual Conference (EMBS 2005), Shanghai, China, 17–18 January 2006; pp. 5881–5883.

17. Pallas-Areny, R. On the reduction of interference due to common mode voltage in two-electrode biopotential amplifiers. *IEEE Trans. Biomed. Eng.* **1986**, *33*, 1043–1046. [CrossRef] [PubMed]

18. Winter, B.B.; Webster, J.G. Reduction of interference due to common mode voltage in biopotential amplifiers. *IEEE Trans. Biomed. Eng.* **1983**, *30*, 58–62. [CrossRef]

19. Lee, K.; Lee, S.; Sim, K.; Kim, K.; Park, K.S. Noise reduction for non-contact electrocardiogram measurement in daily life. In Proceedings of the 36th Annual Computers in Cardiology Conference (CinC 2009), Park City, UT, USA, 13–16 September 2009; pp. 493–496.

20. Guermandi, M.; Scarselli, E.F.; Guerrieri, R. A driving right leg circuit (DgRL) for improved common mode rejection in bio-potential acquisition systems. *IEEE Trans. Biomed. Circuits Syst.* **2016**, *10*, 507–517. [CrossRef]

21. Steffen, M.; Aleksandrowicz, A.; Leonhardt, S. Mobile noncontact monitoring of heart and lung activity. *IEEE Trans. Biomed. Circuits Syst.* **2007**, *1*, 250–257. [CrossRef]

22. Winter, B.B.; Webster, J.G. Driven-right-leg circuit design. *IEEE Trans. Biomed. Eng.* **1983**, *30*, 62–66. [CrossRef]

23. Villegas, A.; McEneaney, D.; Escalona, O. Arm-ECG Wireless Sensor System for Wearable Long-Term Surveillance of Heart Arrhythmias. *Electronics* **2019**, *8*, 1300. [CrossRef]

24. Ding, J.; Tang, Y.; Zhang, L.; Yan, F.; Gu, X.; Wu, R. A novel front-end design for bioelectrical signal wearable acquisition. *IEEE Sens. J.* **2019**, *19*, 8009–8018. [CrossRef]

25. Xu, J.; Mitra, S.; Matsumoto, A.; Patki, S.; VanHoof, C.; Makinwa, K.A.A.; Yazicioglu, R.F. A wearable 8-channel active-electrode EEG/ETI acquisition system for body area networks. *IEEE J. Solid-State Circuits* **2014**, *49*, 2005–2016. [CrossRef]

26. Sakuma, J.; Anzai, D.; Wang, J. Performance of human body communication-based wearable ECG with capacitive coupling electrodes. *Healthc. Technol. Lett.* **2016**, *3*, 222–225. [CrossRef] [PubMed]

27. Serteyn, A.; Vullings, R.; Meftah, M.; Bergmans, J.W.M. Motion artifacts in capacitive ECG measurements: Reducing the combined effect of DC voltages and capacitance changes using an injection signal. *IEEE Trans. Biomed. Eng.* **2015**, *62*, 264–273. [CrossRef] [PubMed]

28. Yoon, S.W.; Min, S.D.; Yun, Y.H.; Lee, S.; Lee, M. Adaptive motion artifacts reduction using 3-axis accelerometer in e-textile ECG measurement system. *J. Med. Syst.* **2008**, *32*, 101–106. [CrossRef] [PubMed]

29. Rodrigues, R.; Couto, P. A neural network approach to ECG denoising. *arXiv* **2012**, arXiv:1212.5217.

30. AAMI; ANSI. *Diagnostic Electrocardiographic Devices*; Standard ANSI/AAMI EC11; AAMI: Washington, DC, USA, 2007.

31. Wu, R.; Tang, Y.; Li, Z.; Zhang, L.; Yan, F. A novel high input impedance front-end for capacitive biopotential measurement. *Med. Biol. Eng. Comput.* **2018**, *56*, 1343–1355. [CrossRef] [PubMed]

*Article*

# New RSA Encryption Mechanism Using One-Time Encryption Keys and Unpredictable Bio-Signal for Wireless Communication Devices


Hoyoung Yu [1] and Youngmin Kim [2],*

[1] Department of Computer Engineering, Kwangwoon University, Seoul 01897, Korea; hoyung134@gmail.com
[2] School of Electronic and Electrical Engineering, Hongik University, Seoul 04066, Korea
* Correspondence: youngmin@hongik.ac.kr; Tel.: +82-2-320-1665


Received: 6 December 2019; Accepted: 28 January 2020; Published: 2 Febuary 2020


**Abstract:** Applying the data encryption method used in conventional personal computers (PC) to wireless communication devices such as IoT is not trivial. Because IoT equipment is extremely slow in transferring data and has a small hardware area compared with PCs, it is difficult to transfer large data and perform complicated operations. In particular, it is difficult to apply the RSA encryption method to wireless communication devices because it guarantees the stability of data encryption because it is difficult to factor extremely large prime numbers. Furthermore, it has become even more difficult to apply the RSA encryption method to IoT devices as a paper recently published indicated that it enables rapid fractional decomposition when using RSA encryption with a prime number generated through several pseudo-random number generators. To compensate for the disadvantages of RSA encryption, we propose a method that significantly reduces the encryption key using a true prime random number generator (TPRNG), which generates a prime number that cannot be predicted through bio-signals, and a disposable RSA encryption key. TPRNG has been verified by the National Institute of Standards and Technology. The NIST test and an RSA algorithm are implemented through Verilog.



**Keywords:** RSA; bio-signal; TRNG; data encryption; IoT; wireless communication


---

## 1. Introduction

The 4th Industrial Revolution is the convergence of information and communications. Data resulting from the 3rd Industrial Revolution are shared among devices through wireless communication and then used and processed for various purposes. A representative technology of the 4th Industrial Revolution is the Internet of Things (IoT). IoT is an infrastructure that enables the communication of objects by information exchange. It can control home devices in certain areas and display all information pertaining to the present situation in real time through a video device such as a camera. As such, human life has become more convenient; however, IoT devices do not guarantee personal privacy when exchanging data through wireless communication. Therefore, data encryption is required during data exchange.

Generally, the data encryption method is divided into symmetric and asymmetric key encryptions. The symmetric key encryption method uses the same secret key for encryption and decryption between the transmitting and receiving sides. To prevent the secret key from being leaked in advance, it must be transmitted through a secure transmission method such as a secret communication network or a direct transmission. The Advanced Encryption Standard (AES) [1] and Data Encryption Standard (DES) [2] are typical symmetric key encryption methods. The symmetric key cryptosystem affords fast encryption and decryption rates, but the data cannot be safely protected even if one of the two sides of the transmitter side is leaked. In addition, the symmetric key cryptosystem is inefficient in many-to-many communication methods such as IoT because the key of each device must be separately generated when a large number of devices is connected [3].

Asymmetric key cryptography or public key cryptography uses different keys for encryption and decryption. The key for encryption is called the public key, and it can be easily used by anyone. The secret key for decryption is kept in a safe place, accessible only by the receiving end that receives the encrypted data. Even if encrypted data are acquired, it is extremely difficult to decrypt them if the private key is not known. Typical asymmetric key cryptosystems are RSA [4] and ECC [5]. In an embedded system such as IoT, power must be used efficiently. The RSA encryption method used in conventional PCs cannot cope with the memory space and power consumption for calculation, as it uses an extremely large key. Therefore, although the ECC encryption method has been developed, the computation process is highly complicated, and it is limited to algorithm expansion because it is developed to use only elliptic curves.

Generally, IoT primarily uses the ZigBee [6] or WiFi [7] communication method; similar to most wireless communication methods, security problems exist in ZigBee communication. High-performance wireless communication methods such as WiFi solve this problem through a high-level encryption process; however, it is difficult to apply high-level encryption technology to ZigBee because of the lack of performance of the terminal itself. The existing ZigBee communication uses the AES-CCM [8] method for data encryption; however, the AES-CCM method does not guarantee confidentiality if the key is leaked even if by only one device through the key transmission process or various methods.

In the one-to-one communication method, the asymmetric key encryption method is primarily used, and in the one-to-many or many-to-many communication method, the symmetric key encryption method is effective. Therefore, the asymmetric key encryption method is effective in IoT equipment, which is a communication method. However, it is difficult to apply RSA encryption to wireless communication or small devices because it requires an extremely large encryption key for security.

Hence, the existing encryption methods used in PCs are not suitable to be used in devices that exchange data in real time. Further, because information regarding temperature, time, etc., does not require high security, it is inefficient to use the existing encryption method as it is. To secure the security of the existing RSA encryption, a key size of 2048 bits is generally required. Transmission of 2048 bits through wireless communication such as ZigBee consumes considerable power, and arithmetic computation with such a large bit number is expensive. Therefore, it is impossible to use the existing RSA's 2048 bit key. In addition, memory is required to store the 2048 bit private key, and preparation for various side attacks is required because it has a memory to save the key.

We herein propose a feasible cryptosystem for situations where extremely high security is not required but real-time encrypted data must be exchanged. It introduces a new encryption mechanism based on the RSA algorithm with a small key and a true prime random number generator for discarding and regenerating keys in real time.

## 2. RSA Overview

The RSA algorithm was first used to implement the concept of public key cryptography and has been widely used because it is easier to understand and implement than other public key algorithms. However, the RSA algorithm is computationally intensive with very large integer numbers. Strong primes are required for RSA security. Thus, additional cost is indispensable for generating strong primes in RSA [9]. The RSA key generation formula is defined as follows:

$$Random\ prime\ number\ select = p,\ q\ (p \neq q)$$
$$n = p \times q$$
$$\varphi(N) = (p-1)(q-1) \tag{1}$$
$$(\varphi(N), e) = 1 < e < \varphi(N)$$
$$Calculate\ d \rightarrow e \times d\ mod\ \varphi(N) = 1$$

For encryption, $e$ is made available to anyone. Furthermore, $d$ for decryption is disclosed only to users requiring decryption. Therefore, $e$ may be leaked, but $d$ should not be leaked. $P$ is the original message, and $C$ is the encrypted message. The process of encrypting and decrypting data is defined as follows:

$$C = P^e\ mod\ n$$
$$P = C^d\ mod\ n \tag{2}$$

The first step of the RSA algorithm is to generate two prime numbers, $p$ and $q$, through a random number generator. The values of $p$ and $q$ should be generated as unpredictable random numbers of different values. The generated $p$ and $q$ are used to calculate $\varphi(N)$ and $n$, respectively. Furthermore, $\varphi(N)$ is used to calculate $e$. $e$ is used for text encryption, and $d$ can be obtained via $e$. $d$ is used to decrypt the encrypted text and must be kept secure such that only the user can view it. If $d$ is leaked, the RSA algorithm is completely broken.

### 3. True Prime Random Number Generator

A prime number generator using a pseudo random number generator (PRNG) causes a fatal defect in RSA security. Recently, a method for quickly obtaining the private key of the RSA algorithm using PRNG has been reported [10]. Therefore, to maintain RSA security, an unpredictable true random prime number generator (TRPNG) is required. A TRNG with high entropy is required to generate random numbers that cannot be predicted. In this study, we use TRNG based on Linear-feedback shift register (LFSR) using bio-signals [11]. Photoplethysmogram (PPG) sensors [12], which are built into most wearable devices, measure values by capturing changes in the arterial perfusion rate of light as the arterial blood flow varies with each heartbeat. The PPG sensor suffers from noise due to physical and environmental factors such as light entering from the outside when moving a finger or arm. However, this disadvantage is an advantage when implementing the TRNG. This is because the value is changed every time according to the fine movement of the person or the surrounding environment, which cannot be predicted.

We use the most representative 16 bit LFSR of the PRNG because PPG data alone cannot generate the same ratio of zero and one, which are the most important factors in generating random numbers. The fact that zero and one are equal implies that the maximum length sequence of the generated bits is output, and the polynomial is set to $x^{16} + x^{15} + x^{13} + x^{14} + 1$ [13]. In this study, we use the initial seed value as a physical source obtained from the PPG sensor because the LFSR requires an initial seed value for random number generation. Because the physical sources that can be obtained from PPG sensors are generally processed in wearable devices, no additional processing is required. When the 16 bit LFSR outputs the initial seed value and 65,535 random numbers, which is the maximum length sequence of 16 bits, a problem arises in that the random number of the same pattern is repeated, as shown in Figure 1. Hence, a new polynomial is added. An XOR gate is added, as shown in Figure 2, to perform the XOR operation with the physical source value of the PPG sensor, in which the random number generated in the LFSR cannot be predicted.

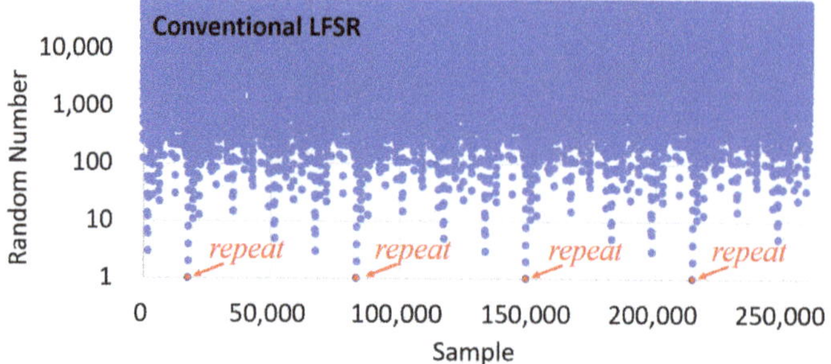


**Figure 1.** Conventional LFSR random number pattern.

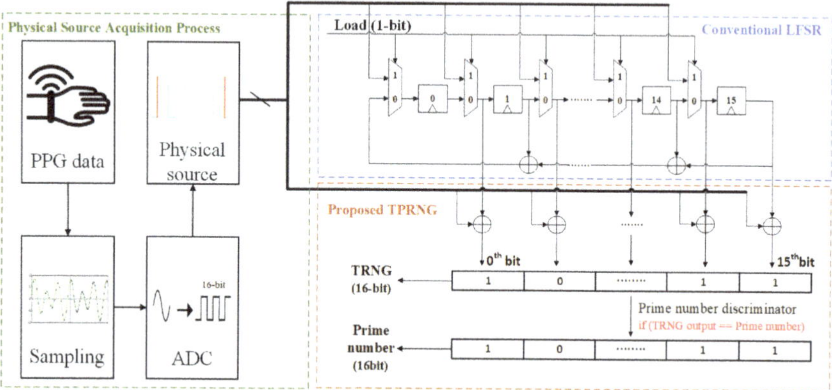


**Figure 2.** Proposed true prime number generator block diagram.

To use the random numbers generated by the TRNG as the $p$ and $q$ values of the RSA algorithm, a discriminator is required to determine the prime number. Because a discriminator through the factorial decomposition requires considerable time and hardware area to discriminate the prime numbers, the random numbers generated from the TRNG are compared with the decimal values stored as the parameter values to be determined as a prime number. If the random number generated by the TRNG is equal to the decimal value stored in the parameter, it is determined to be a prime number, and the output random prime is $p, q$ in the RSA algorithm. The reason for using unpredictable prime numbers as $p$ and $q$ values is that the prime factorization algorithm from a recently published paper [10] cannot be applied.

## 4. Proposed RSA

The RSA algorithm requires a key of 2048 bits or more to guarantee security. The encryption algorithm using such a large key size is not suitable for use in wireless communication devices, small devices, or places requiring fast data processing. Therefore, the RSA algorithm is not used in IoT devices or cell phones, which constantly exchange data and are being miniaturized increasingly. AES encryption, a symmetric key encryption method, affords an extremely fast processing speed and a small hardware area for encrypting and decrypting data; however, it is inefficient in one-to-many or many-to-many communications. Further, even if one device key is leaked, the security of all devices cannot be maintained. The use of symmetric keys is risky because IoT devices communicate with many other devices. ECC exploits the idea that it takes a long time to find a discrete log of a random elliptic curve for a particular known point [5]. As shown in Table 1 [14], ECC encryption can provide a similar level of security while using a key with a much shorter length than that of the RSA. However, the elliptic curve is complicated and expensive to operate.

**Table 1.** Key size on equivalent strength between RSA and ECC.

| Time to Break in MIPS Years | RSA Key Size | ECC KEY Size | RSA Key Size Ratio |
| --- | --- | --- | --- |
| $10^4$ | 512 | 106 | 5:1 |
| $10^8$ | 768 | 132 | 6:1 |
| $10^{11}$ | 1024 | 160 | 7:1 |
| $10^{20}$ | 2048 | 210 | 10:1 |
| $10^{78}$ | 21,000 | 600 | 35:1 |

As shown in Figure 3, the existing RSA encryption is primarily classified into three processing modules. First, the key generator module receives a 1024 bit pseudo-random number and generates both a private and public key. The second storage module is a memory that holds a 2048 bit private key. It is highly vulnerable to side attacks because it stores a 2048 bit key in memory continuously. If the side attack is successful, the security of the RSA encryption algorithm is lost regardless of the encryption key size. Next, the security module uses 2048 bit private and public keys to encrypt data. The algorithm for encrypting and decrypting through the 2048 bit key requires extremely large hardware, which is virtually impossible to implement. In addition, it is difficult to exchange 2048 bit private and public keys through wireless communication.

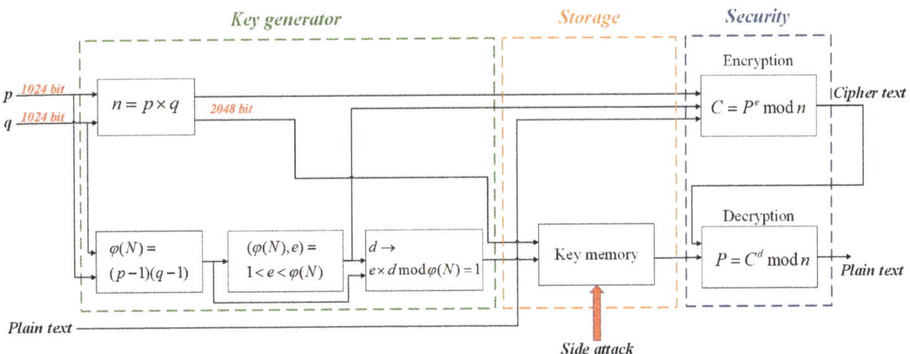


**Figure 3.** Conventional RSA block diagram.

To solve the problem of the conventional RSA, we herein propose a new RSA mechanism, as shown in Figure 4. The proposed RSA mechanism obtains two 16 bit random prime numbers in the TPRNG using unpredictable PPG data. The next two 16 bit prime numbers $p$ and $q$ are used to generate 32 bit public and private keys. When both a public and private key are generated, the public key $(e, n)$ is distributed to the device requiring encryption and the plain text is encrypted. Encrypted cipher text is sent where data are required and decrypted via private key $(d)$. When all the steps from key generation to decryption are completed, the prime number, public key, and private key are immediately destroyed and then regenerated. The RSA contains a key that is much smaller than the key used by the RSA, which can be deciphered at a much faster rate than the existing 2048 bit key; however, it can be estimated by constantly regenerating the key. In addition, real-time transmission is possible when transmitting key and encrypted data through wireless communication because it has extremely small keys. In terms of hardware design for encryption and decryption, the side attack does not apply because of the destruction and regeneration through the process, rather than maintaining the key in memory constantly.

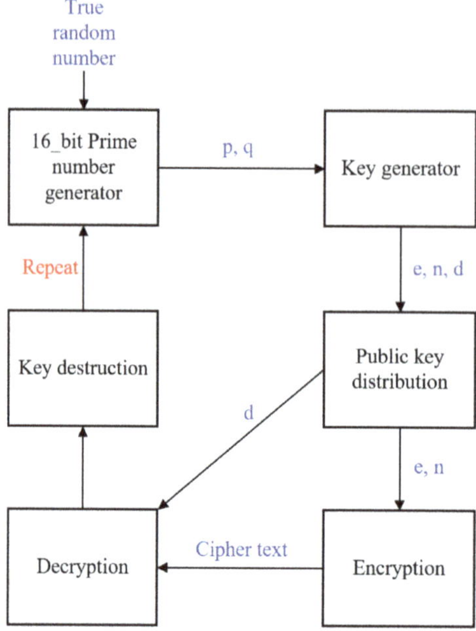

**Figure 4.** Proposed RSA encryption mechanism.

## 5. Implementation

For the proposed hardware design, we used the Verilog HDL, FPGA in Zynq UltraScale$^+$, xczu6cg-ffvb1156-2 device. This design was synthesized and implemented using the Vivado design suite provided by Xilinx. NIST SP 800-22 was used to evaluate the randomness of the implemented TRNG.

### 5.1. TPRNG

An unpredictable random number can be generated as shown in Figure 5 by performing an XOR operation with the random number generated in the conventional LFSR and the physical source value of the PPG sensor, as shown in Figure 1. The proposed random number generator is verified by all passing the NIST test suite [15], as shown in Table 2. As explained in Section 3, to implement TPRNG, the random numbers generated through the TRNG are compared with the numbers stored in the parameters and then output as a decimal number. If the number of decimals stored as the parameter value is extremely small, the decimal number is assessed whether it is real time. Thus, a decimal value of 1000 or less is not stored as the parameter value. The TPRNG has an extremely small hardware area, as shown in Table 3.

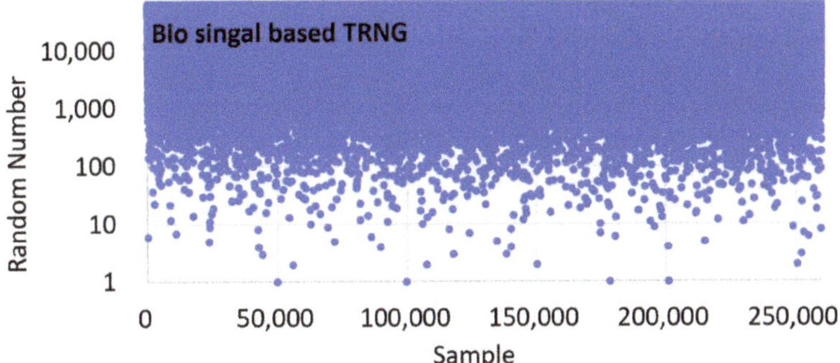


**Figure 5.** Random number pattern generated in true random number generator.

**Table 2.** NIST test suite result. P: pass, F: fail.

| Parameters | Conventional | Proposed |
|---|---|---|
| Frequency | P | P |
| Block Frequency | P | P |
| Cumulative Sums | F | P |
| Runs | F | P |
| Longest Run | F | P |
| Non-Overlapping Template | F | P |
| Overlapping Template | F | P |
| Approximate Entropy | F | P |
| Random Excursions | F | P |
| Random Excursions Variant | F | P |
| Serial | F | P |
| Linear Complexity | P | P |

**Table 3.** True prime random number generator (TPRNG) resource usage.

| Site Type | Used | Available |
|---|---|---|
| LUTs | 188 | 214,604 |
| Registers | 86 | 429,208 |
| BRAM | 0 | 714 |
| DSP | 0 | 1973 |

*5.2. RSA*

We herein propose an RSA system architecture, as shown in Figure 4. The two prime numbers generated by the TPRNG are $p$, $q$ with the RSA. The private key ($d$) and public key ($e$) are computed using the $p$ and $q$ values, respectively. Data are encrypted and decrypted through the generated $e$ and $d$, respectively. Once all the encryption and decryption processes have been completed through $e$ and $d$, the TPRNG is used to regenerate the key by inputting new $p$ and $q$ values. When the RSA algorithm is implemented in hardware, the hardware area increases exponentially, as shown in Figure 6, as the number of bits increases. The elements used for each bit are as shown in Table 4. Therefore, a 32 bit

RSA encryption is suitable for IoT devices requiring small hardware area and low power. The RSA encryption module implemented in hardware is shown in Figure 7 [16–18]. The $p, q$ values and plain text are input in the decryption module, and the keys ($n, e, d$) are generated according to the RSA algorithm. The 1542 value of the plain text according to the RSA encryption formula $C = P^e \ mod \ n$ is encrypted to the 8,135,351 value cipher text. The decryption module implemented in hardware is shown in Figure 8. Cipher text is transmitted to the input of the decryption module and decrypted to the 1542 value plain text through the 3,849,029 value private key ($d$) according to the RSA decryption formula $P = C^d \ mod \ n$. Once the decryption is completed, the encryption and decryption keys are discarded.

**Table 4.** RSA implementation size by each bit.

|           | **32 bit** | **64 bit** | **128 bit** | **256 bit** |
|-----------|-----------|-----------|------------|------------|
| LUTs      | 6367      | 25,139    | 100,375    | 412,391    |
| Registers | 301       | 741       | 2234       | 5310       |
| DSP       | 13        | 44        | 148        | 964        |

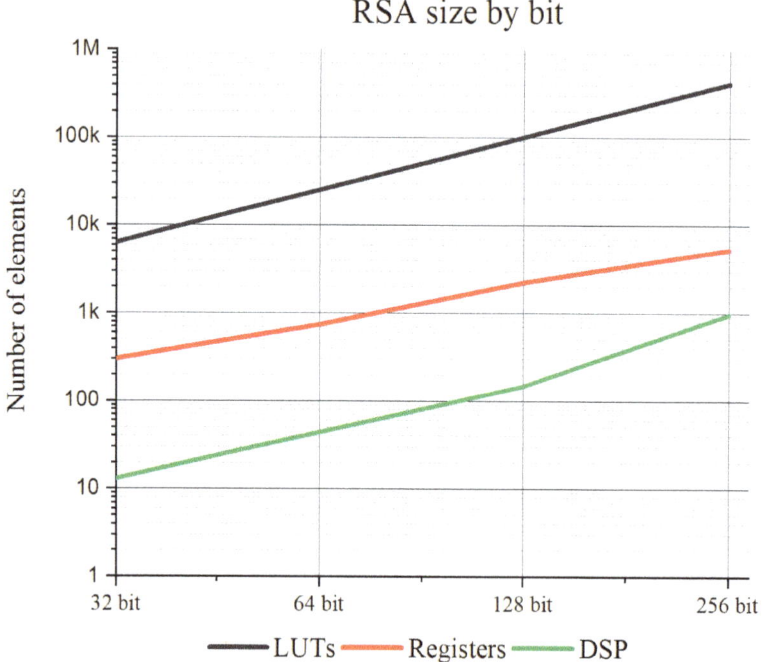


**Figure 6.** RSA size by bit length.

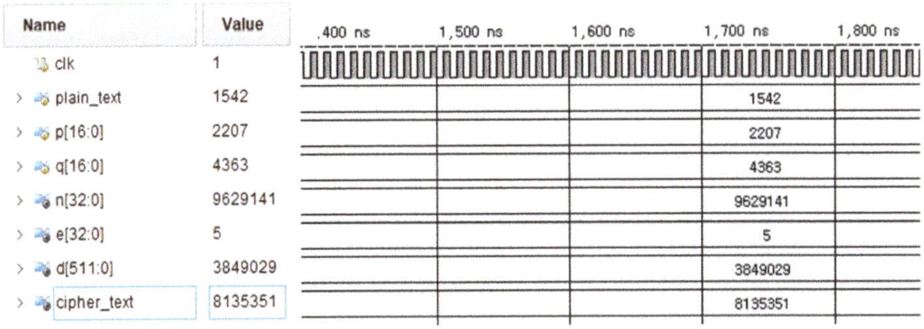

Figure 7. Encryption simulation results.

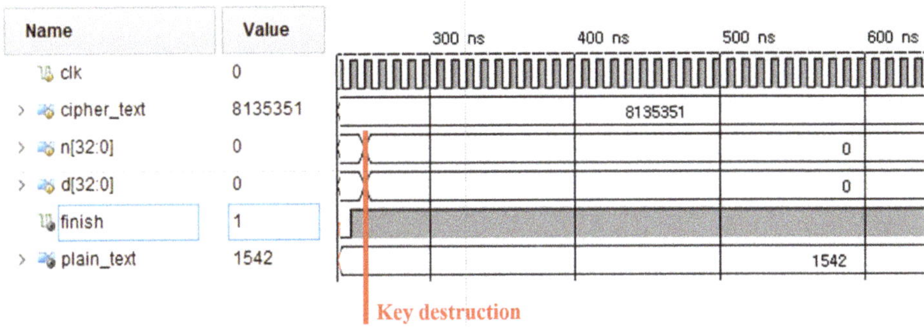


**Figure 8.** Decryption simulation results.

## 6. Conclusions

We herein proposed a new RSA mechanism applicable to small wireless communications devices such as IoT. Conventional RSA algorithms use a very large key size, which requires large hardware areas and expensive arithmetic calculations. For this reason, a traditional RSA encryption is not suitable for devices used in IoT environments. The proposed RSA encryption mechanism has an extremely small key, but compensates for the problem of small encryption keys by continuously changing the key using an unexpected random number. Furthermore, side attack problems do not occur because the key does not remain in memory continuously. The proposed method is highly suitable for IoT devices that use many-to-many communications, as it does not suffer from problems in existing public key encryptions, which hinder data encryption. Further, as the information exchanged between IoT devices does not require high security, this method is feasible in that data are decrypted quickly with little hardware. The RSA mechanism proposed in this paper exploits the randomness of the bio-signals with a very small number of keys (e.g., 16 bits) for power and area efficiency in RSA encryption. For this reason, the proposed method may not be sufficient for environments that require an extremely high level of encryption. Future work will evolve into research that increases area and power efficiency even with greater key size. This method is applicable to cases where data must be transmitted and received quickly in real time, e.g., an unmanned vehicle.

**Author Contributions:** Conceptualization, H.Y. and Y.K.; methodology, H.Y.; software, H.Y.; validation, H.Y. and Y.K.; investigation, H.Y.; resources, H.Y. and Y.K.; writing, original draft preparation, H.Y.; writing, review and editing, Y.K.; supervision, Y.K.; project administration, Y.K.; funding acquisition, Y.K. All authors have read and agreed to the published version of the manuscript.

**Funding:** This research was funded by MIST under Grant NRF-2019M3F3A1A02072093.

**Acknowledgments:** This work was supported by the NRF of Korea funded by the MSIT under Grant NRF-2019M3F3A1A02072093 (Intelligent Semiconductor Technology Development Program).

**Conflicts of Interest:** The authors declare no conflict of interest.

## References


1. NIST. *Advanced Encryption Standard*; FIPS Publication 197; NIST: Gaithersburg, MD, USA, 2001.
2. Eli, B.; Shamir, A. Differential cryptanalysis of DES-like cryptosystems. *J. Cryptol.* **1991**, *4*, 3–72.
3. INFOSEC. Available online: https://resources.infosecinstitute.com/review-asymmetric-cryptography/#gref/ (accessed on 6 November 2019).
4. Rivers, R.L.; Shamir, A.; Adleman, L. A method for obtaining digital signatures and public-key cryptosystems. *Commun. ACM* **1978**, *21*, 120–126.
5. Singh, L.D.; Singh, K.M. Implementation of text encryption using elliptic curve cryptography. *Procedia Comput. Sci.* **2015**, *54*, 73–82. [CrossRef]
6. ZigBee, A. Available online: https://www.eena.org/ (accessed on 6 November 2019).
7. Beal, V. Available online: www.webopedia.com/TERM/W/Wi_Fi.html (accessed on 6 November 2019).
8. Vidgren, N.; Hataja, K.; Patiño-Andres, J.L.; Ramírez-Sanchis, J.J.; Toivanen, P. Security Threats in ZigBee-Enabled Systems: Vulnerability Evaluation, Practical Experiments, Countermeasures, and Lessons Learned. In Proceedings of the 2013 46th Hawaii International Conference on System Sciences, Maui, HI, USA, 7–10 January 2013.
9. Rivest, R.; Silverman, R. Are 'Strong' Primes Needed for RSA. In *Cryptology ePrint Archive, Report 2001/007.* Available online: https://eprint.iacr.org/2001/007 (accessed on 6 November 2019).
10. Nemec, M.; Sys, M.; Svenda, P.; Klinec, D.; Matyas, V. The Return of Coppersmith's Attack: Practical Factorization of Widely Used RSA Moduli. In Proceedings of the 2017 ACM SIGSAC Conference on Computer and Communications Security, Dallas, TX, USA, 30 October–3 November, 2017; pp. 1631–1648.
11. Yu, H.; Kim, Y. True Random Number Generator Using Bio-related Signals in Wearable Devices. In Proceedings of the 2018 Conference on International SoC Design Conference, Daegu, Korea, 12–15 November 2018.
12. Wood, P.T.; Wood, L.B.; Asada, H.H. Active motion artifact cancellation for wearable health monitoring sensors using collocated MEMS accelerometers. In *Smart Structures and Materials 2005: Sensors and Smart Structures Technologies for Civil, Mechanical, and Aerospace Systems, International Society for Optics and Photonics*; International Society for Optics and Photonics: San Diego, CA, USA, 2005; Volume 5765, pp. 811–820.
13. Hellebrand, S. Built-in test for circuits with scan based on reseeding of multiple-polynomial linear feedback shift registers. *IEEE Trans. Comput.* **1995**, *44*, 223–233. [CrossRef]
14. Yu, Z. A High Performance Pseudo-Multi-Core Elliptic Curve Cryptographic Processor over GF ($2^{163}$). Ph.D. Thesis, University of Saskatchewan, Saskatoon, SK, Canada, 2010.
15. Rukhin, A. *A Statistical Test Suite for Random and Pseudorandom Number Generators for Cryptographic Applications*; NIST: Gaithersburg, MD, USA, 2010.
16. Shams, R.; Khan, F.H.; Umair, M. Cryptosystem an Implementation of RSA Using Verilog. *Int. J. Comput. Netw. Commun. Security* **2013**, *1*, 102–109.
17. Rahman, M.; Rokon, I.R.; Rahman, M. Efficient Hardware Implementation of RSA Cryptography. In Proceedings of the 3rd International Conference on Anti-Counterfeiting, Security, and Identification in Communication, Hong Kong, China, 20–22 August 2009; pp. 316–319.
18. Leelavathi, G.; Shaila, K.; Venugopal, K.R. Design and Implementation of Montgomery Multipliers in RSA Cryptography for Wireless Sensor Networks. In *Proceedings of First International Conference on Smart System, Innovations and Computing*; Springer: Singapore, 2018; pp. 565–574.

*Article*

# Development of a Compact, IoT-Enabled Electronic Nose for Breath Analysis

Akira Tiele, Alfian Wicaksono, Sai Kiran Ayyala and James A. Covington *

School of Engineering, University of Warwick, Coventry CV4 7AL, UK; F-A.Tiele@warwick.ac.uk (A.T.);
A.Wicaksono@warwick.ac.uk (A.W.); Sai-Kiran.Ayyala@warwick.ac.uk (S.K.A.)
* Correspondence: J.A.Covington@warwick.ac.uk; Tel.: +44-(0)-24-7657-4494

Received: 29 November 2019; Accepted: 28 December 2019; Published: 1 January 2020

**Abstract:** In this paper, we report on an in-house developed electronic nose (E-nose) for use with breath analysis. The unit consists of an array of 10 micro-electro-mechanical systems (MEMS) metal oxide (MOX) gas sensors produced by seven manufacturers. Breath sampling of end-tidal breath is achieved using a heated sample tube, capable of monitoring sampling-related parameters, such as carbon dioxide ($CO_2$), humidity, and temperature. A simple mobile app was developed to receive real-time data from the device, using Wi-Fi communication. The system has been tested using chemical standards and exhaled breath samples from healthy volunteers, before and after taking a peppermint capsule. Results from chemical testing indicate that we can separate chemical standards (acetone, isopropanol and 1-propanol) and different concentrations of isobutylene. The analysis of exhaled breath samples demonstrate that we can distinguish between pre- and post-consumption of peppermint capsules; area under the curve (AUC): 0.81, sensitivity: 0.83 (0.59–0.96), specificity: 0.72 (0.47–0.90), $p$-value: <0.001. The functionality of the developed device has been demonstrated with the testing of chemical standards and a simplified breath study using peppermint capsules. It is our intention to deploy this system in a UK hospital in an upcoming breath research study.

**Keywords:** breath analysis; electronic nose (E-nose); Internet-of-Things (IoT)

---

## 1. Introduction

The diagnostic potential of breath was first utilised by ancient Greek physicians who understood that distinct odours and aromas could be related to specific diseases [1]. In recent years, applications of breath analysis have focused on the non-invasive detection of volatile organic compounds (VOCs), using different technological platforms. It has been suggested that there are over 3000 VOCs in human breath that are a combination of by-products of normal metabolic activity and, in some cases, specifically associated with a disease [2]. Changes in VOC composition or concentrations serve as potential biomarkers for the detection and monitoring of disease, such as chronic obstructive pulmonary disease (COPD) [3], colorectal cancer [4], and many more [5]. Gas chromatography-mass spectrometry (GC-MS) is generally considered the "gold standard" for breath research and is the most commonly used technique [6]. GC-MS is considered an offline method because samples need to be collected (and potentially stored) onto, for example, thermal desorption (TD) tubes or sampling bags, prior to analysis. Other commonly used technologies include proton-transfer-reaction mass-spectrometry (PTR-MS) and selected-ion flow tube mass spectrometry (SIFT-MS) [5]. These technologies are on-line methods (suitable for real-time analysis) and have been used for both disease diagnosis studies [7] and monitoring pharmacokinetics effects [8]. Although these analytical platforms are highly reproducible and accurate, they are very expensive, time-consuming, and lack portability.

To fully utilise the diagnostic potential of exhaled breath, Hunter and Dweik [9] argued that its application must be extended beyond laboratories and pilot studies to standard clinical practice and at home. The latter requires a compact, personal and portable diagnostic device, capable of sampling and analysing the breath of an individual at any time or place. The authors referred to this as a "smart breath healthy diagnostic system" (SBHDS). Since this concept was first proposed in 2008, the technological revolution of Internet-of-things (IoT) has invigorated the demand for smarter sensor solutions. This also applies to the medical and healthcare domain (so-called IoMT: internet of medical things), which seeks to integrate medical devices into IoT networks. Personalised tools for health monitoring could reduce the overall costs of care and bring numerous benefits to health professionals and patients [10]. Utilising the computing power of smartphones, these sensor systems could analyse and transmit collected data from a patient to a hospital cloud computing-based framework. We believe that personalised breath analysis devices for non-invasive diagnosis and/or monitoring of disease have potential for integration into this future IoMT-framework. However, the currently used technologies are unable to fulfil the requirements of such a device.

While no single VOC analysis technology can provide the complete diagnosis of an individual, the electronic nose (E-nose) could be considered a good option as it has significant advantages. These include low-cost, low-power (compared to GC-MS), user-friendliness, and portability. An 'E-nose' describes an instrument consisting of an array of cross-reactive gas sensors, coupled with pattern recognition software. It relies on each sensor in the array being different and therefore its response to an odour being unique within the array. The pattern recognition software then learns the sensor responses associated with a specific odour source [11]. This operating principle attempts to mimic the function of biological olfactory receptors by detecting a complex pattern, instead of measuring the individual constituents of a mixture. Gas sensors suitable for E-noses can be broadly divided into four main types, namely resistive, catalytic, optical, and electrochemical [12]. Resistive gas sensors are often based on semiconductor metal oxides that have been widely used for E-noses, due to their fast response/recovery time, low-cost, and sensitivity to a wide range of target gases [13]. In recent years, the application of micro-electro-mechanical systems (MEMS) technology has allowed metal oxide (MOX) gas sensors to dramatically reduce in size and power consumption. Mass-production has allowed these sensors to be produced at even lower cost. These developments provide new opportunities in medical diagnostic applications, such as SBHDS-inspired E-noses.

In a 2018 review paper [14] regarding the applications of E-noses, at least 50 papers related to non-invasive detection of human diseases; 27 of which analysed exhaled breath. The commercial E-nose instrument most frequently used (14/27) for breath was Cyranose 320 (Sensigent, Baldwin Park, CA, USA). This E-nose contains an array of 32 carbon black polymer composite (CBPC) sensors. The advantage of CBPC sensors (compared to MOX) is that they do not require high temperature operation and therefore have lower power consumption. The disadvantage of this sensing technology is that they demonstrate poor repeatability and reproducibility, due to the random nature of the polymer, and have a relatively short sensor life, compared to MOX sensors [15]. Other sensor technologies included in the review were quartz crystal microbalance (QMB), nitric oxide sensor (NOS), gold nanoparticle (GNP), and carbon nanotubes (CNT). Only three papers in the review used E-noses with MOX sensors [16–18]. In a different review paper, regarding the use of E-nose technology for the diagnosis of digestive and respiratory diseases using exhaled breath, MOX sensor-based E-noses are similarly under-represented (four of 23 reviewed papers) [19]. These reviews demonstrate that MOX sensors are currently under-utilised in E-nose applications for exhaled breath research. While new prototype MOX-based E-noses are currently being developed for diagnostic purposes using exhaled breath [20], these often use custom sensors or traditional commercial MOX sensors/E-noses, which have been available for many years [21]. The latest generation of commercially available MEMS MOX gas sensors have not been sufficiently evaluated for this purpose.

Some researchers have used a selection of these sensors in devices for air quality monitoring purposes [22,23], but applications in breath research are limited. The breath E-nose developed by Jaeschke et al. [24] uses many of the most relevant commercial gas sensors currently available. However, this unit focuses on a modular approach with three exchangeable sensing compartments and has not been tested using exhaled breath samples. This provides an opportunity to utilise MEMS MOX gas sensors and existing IoT platforms to develop a new generation of breath analysis E-noses.

In our previous work [25], we developed a E-nose which was referred to as WOLF (Warwick OLFaction). This system utilised 13 gas sensors and multiple sensor technologies (10 electrochemical, two optical and one photo-ionisation detector) and was intended for head-space gas analysis. In this work, we report on the development of the latest in-house built WOLF E-nose, designed specifically for breath analysis: WOLF Breath E-nose.

## 2. Materials and Methods

### 2.1. Breath Sampling

The developed unit comprises of 2 sub-systems: breath sampling and breath analysis. The breath sampling system is made up of a 16 cm long, 2 cm diameter, aluminium tube with 3D printed tube connectors on either end. These components serve as an interface between the sampling tube and the enclosure and were printed on a Form 2 printer using F2GPWH-04 material (Formlabs, Somerville, MA, USA). The front-end connector also acts as a holder for disposable one-way valve mouthpieces (6020-1, Medacx, Hayling Island, UK). The back-end connector was designed with an opening to embed a sensor module (SCD30, Sensirion, Stäfa, Switzerland). The assembly of these components is shown in Figure 1, designed using SolidWorks (ver. 2018, Dassault Systèmes, Vélizy-Villacoublay, France). The sensor module was fitted at the end of the sampling tube to monitor sampling-related parameters, such as carbon dioxide ($CO_2$), humidity, and temperature. Tracking these parameters allows the subject (at home) or operator (in a clinical setting) to check the quality and repeatability of samples.

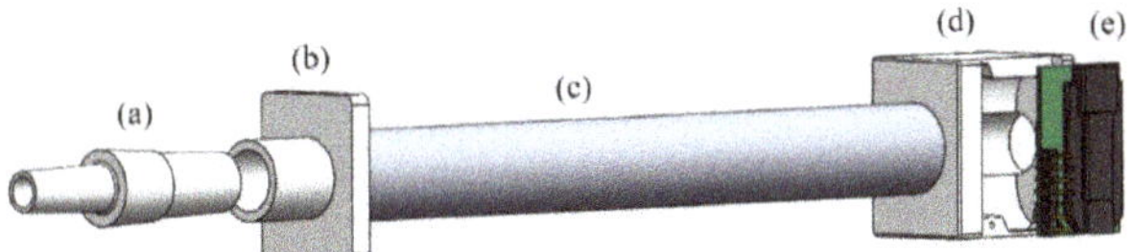


**Figure 1.** 3D model of sampling tube components: (**a**) One-way mouthpiece; (**b**) Front-end tube connector; (**c**) Sampling tube; (**d**) Back-end tube connector; (**e**) Sensor module.

The sampling tube has a volume of approximately 50 mL and can be heated to body temperature (35–37 °C) using a low-cost digital thermostat and heater relay module (W1209, HiLetgo, Shenzhen, China). The heater relay is connected to nichrome (NiCr) wire, which is coiled around the sampling tube. The tube was then wrapped in high temperature resistant Kapton tape (436-2778, RS, Corby, UK) and silicone thermal interface sheets (446-493, RS, Corby, UK) to provide some insulation. The heating procedure is initiated at start-up and takes around 10 min. Heating the sampling tube avoids condensation from forming inside, which could lead to cross-contamination from previous samples.

The integrated sampling system replicates that of the Bio-VOC breath sampler (C-BIO01, Markes Intl., Llantrisant, UK). The Bio-VOC is a commercial breath sampling kit, consisting of an open-ended hard plastic sampling tube. It has been used frequently in recent breath analysis studies [26] for the sampling of end-tidal breath. This refers to the last portion of exhaled air, which has undergone gaseous exchange with blood [27]. As the subject breathes through the tube, air is displaced, which separates dead-space gas from end-tidal breath. This displacement principle has been replicated in our sampling design. The use of one-way valve mouthpieces and heating the sampling tube further improve on the design concept of the Bio-VOC.

## 2.2. Breath Analysis

The breath analysis system includes a custom PCB and sensor chamber. The PCB was designed using Altium Designer (ver. 19.1.6, Altium, Chatswood, Australia). The dimensions of the PCB are 13 × 16 cm. The design has an 8 × 2.5 cm sensing area, which includes an array of 10 MEMS MOX-based gas sensors. These include both thick film and thin film sensors and the AlphaSense dual sensor is the only true 'p-type' material [28]. Both analogue and digital sensors were used. According to the respective datasheets, most of the sensors provide 'total VOC' (TVOC) readings, which produce a single measurement to represent a mixture of VOCs. The total cost of the sensors is around £200. The deployed sensors are summarised in Table 1.

**Table 1.** Sensors deployed in WOLF Breath E-nose.

| Interface | Sensor | Manufacturer | Target Gas |
|---|---|---|---|
| Digital | CCS811 | ams | TVOC |
| | SGP30 | Sensirion | $H_2$, ethanol |
| | BME680 | Bosch | TVOC |
| | iAQ-Core C | ams | TVOC |
| | ZMOD4410 | IDT | TVOC |
| Analog | MiCS-6814 | SGX | $NH_3$, reducing, oxidising |
| | Dual Sensor | AlphaSense | Reducing and oxidising |
| | TGS-8100 | Figaro | TVOC |
| | TGS-2620 | Figaro | TVOC |
| | AS-MLV-P2 | ams | Reducing and CO |

The sensing area is enclosed by a 3D printed sensor chamber (F2GPWH-04/Form 2, Formlabs, Somerville, MA, USA), with dimensions of 4 × 9.5 × 1.8 cm. The chamber was sealed to the PCB using an O-ring and the compartment volume is around 25 mL (excluding sensors). A miniature pump (PMM1031-NMP015, KNF, Witney, UK) was used to create a negative pressure system, which pulls gas through the sensor chamber. There is a 3-way valve (ETO-3-12, Clippard, Cincinnati, OH, USA) between the sampling tube and sensor chamber. This creates 2 flow-paths to the sensor chamber: ambient air (from the environment) or exhaled breath (from the sampling tube). Push-fit connectors and 1/8" PTFE tubing (06605-27, Cole-Parmer, St Neots, UK) were used to create the internal connections (e.g., valve to sensor chamber). The pump runs continuously (flow rate set to 500 mL/min) and pulls ambient air through the sensor chamber, when the device is in an idle state. The flow rate of the pump is controlled using a simple digital-to-analogue converter (DAC) drive circuit. There is no flow feedback, but the flow rate was measured using a flowmeter (7000 Flowmeter, Ellutia, Ely, UK) and voltage set to the desired flow rate. The switching of the valve is also controlled using a DAC drive circuit. When sampling is activated, using the LED-push button on the front-panel of the unit, the valve is triggered to switch the flow-path to that from the sampling tube. The internal layout of the components was modelled and arranged in SolidWorks prior to manufacture, as shown in Figure 2.

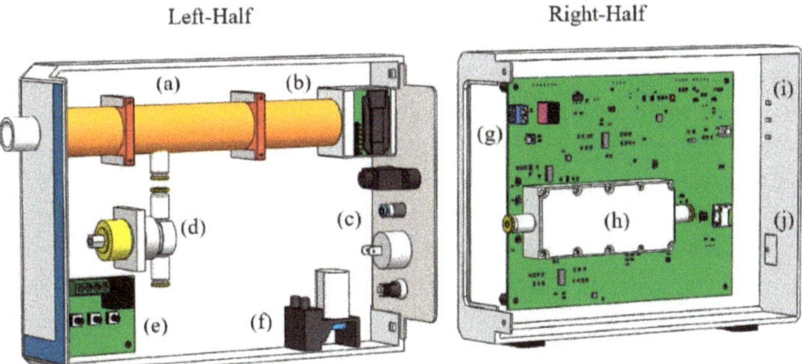


**Figure 2.** Internal layout of WOLF Breath E-nose components: (**a**) Sampling tube; (**b**) Sensor module; (**c**) Back plate with USB, exhaust, on/off button and DC power jack; (**d**) 3-way valve; (**e**) Heater relay module; (**f**) Pump; (**g**) Custom PCB; (**h**) Sensor chamber; (**i**) Indicator LEDs; (**j**) Sampling button.

An internal system view of the WOLF Breath E-nose is shown in Figure 3. The red and blue arrows indicate the different flow-paths through the device. The dimensions of the unit are $7 \times 16 \times 23$ cm and weighs less than 1 kg. An external view of the unit is shown in Figure 4.

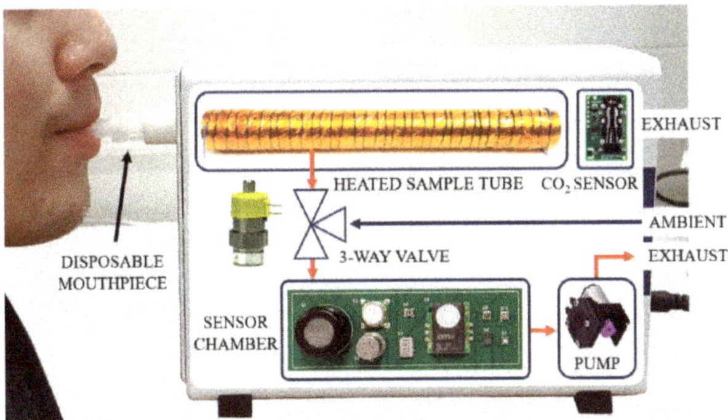


**Figure 3.** Internal system view of the WOLF Breath E-nose.

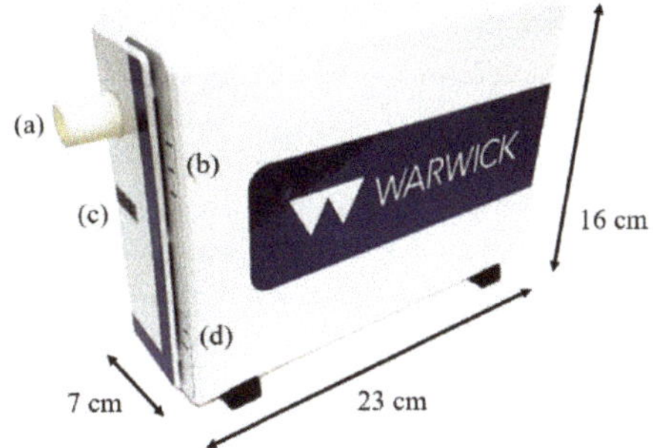


**Figure 4.** WOLF Breath E-nose: (**a**) Mouthpiece holder; (**b**) LED status indicators; (**c**) Sampling tube temperature indicator; (**d**) Sampling button.

The unit can be controlled from a laptop, via wired USB, or using Wi-Fi through a simple mobile app that was developed using Blynk. This app provides a free IoT platform for Android or iOS and is compatible with commonly used IoT hardware, such as Arduino, Raspberry Pi, ESP8266 and

ESP32. The ESP32 (Espressif Systems, Shanghai, China) was used as the microcontroller for our system, because it is a low-cost, low-power chip with integrated Wi-Fi and Bluetooth communication capabilities. The custom 'WOLF Breath E-Nose' app includes 2 tabs: sampling and analysis, as shown in Figure 5. The sampling tab includes 3 input fields to include details such as study name, subject ID and sample name. In addition to this, status indicators from the front-panel of the device are shown, as well as real-time readings of $CO_2$, temperature and humidity from the SCD30 sensor. The parameters can be selected and plotted on the graph, as shown with $CO_2$ in Figure 5 (left). The analysis tab shows the real-time readings from the gas sensor array. Individual or multiple outputs can be selected and plotted on the graph, as shown with CCS811 and BME680 in Figure 5 (middle). The other readings can be seen by scrolling down, as shown in Figure 5 (right). The sampling rate of the sensor data was set to 1 sample per second.

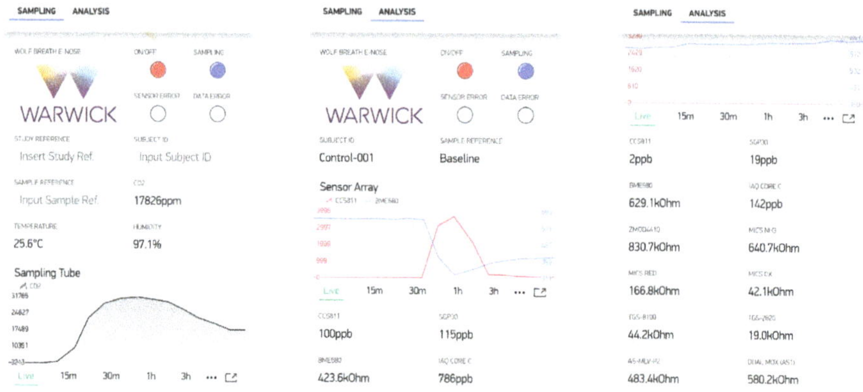

**Figure 5.** WOLF Breath E-nose Blynk app; (**left**) Sampling tab; (**middle**) Analysis tab with real-time output graph; (**right**) Analysis tab with all sensor readings.

## 2.3. Electronic Design

The operating principle of MOX sensors is based on a reduction-oxidation (redox) reaction. In 'clean' air, oxygen is absorbed onto the gas-sensitive layer of the sensor (e.g., tin oxide). When exposed to a reducing or oxidising gas, the oxygen reacts to the target gas, which changes the conductivity of the sensitive layer, resulting in a change in current [29]. The applied voltage is fixed, to measure the change in resistance to characterise the sensor response. Traditionally, the limitations of MOX gas sensors relate to high power consumption and temporal drift [30,31]. The emergence of MEMS-technology MOX sensors has led to further miniaturisation of sensing chips and heater resistors, which has drastically reduced power consumption. In general, thick film MOX sensors are more stable than MEMS (thin film) devices, however MEMS sensors have faster response times [32]. To further reduce power consumption, duty cycling using two temperature pulsing can be used. This involves switching between a higher and lower temperature, to reduce baseline drift, decrease response time and power [33]. For this application, a heating phase of 150 ms was implemented for the analogue MEMS MOX gas sensors.

The power system of the E-nose is currently based around the input from a 12 V AC/DC mains power supply unit (709-GEM30I12-P1J, Mouser, High Wycombe, UK). The system was designed around 12 V, because this is the required voltage for the valve and heater relay module. However, in future, this device may be re-configured to be battery-operated. In addition to the improved sensor stability, the possibility of utilising duty-cycle mode is likely to be beneficial for reducing power consumption when battery-powered.

The schematic shown in Figure 6 shows a simplified circuit diagram of sensor and heater drive circuits for analogue MOX sensors. The 12-bit resolution DAC in the sensor drive circuit dictates the sensor voltage (using a constant voltage configuration) and the previously referred to duty-cycle. The second stage of the sensor drive circuit removes the offset of the voltage bias and amplifies the sensor response, in this case with a gain of 2.2. The sensor output is then read using a 16-bit resolution analogue-to-digital converter (ADC). The heater drive circuit is controlled by a DAC and provides a fixed heater voltage to the heater of the gas sensor. The heater resistance of the sensor can be calculated by measuring the voltage drop, using an ADC, across the heater load resistor. Table 2 provides a summary of the sensor and heater voltages and resistances for deployed analogue gas sensors. The values in Table 2 relating to sensor resistances represent the real resistances of the sensors used in the unit. The heater resistances are those provided by the manufacturer. The values of the sensor and heater load resistors were optimised to allow for the measurement of the widest range of $R_S/R_0$, according to their respective datasheets, where $R_S$ is the resistance of the sensor depending on the concentration of target gas(es) and $R_0$ is the resistance of the sensor at ambient air.

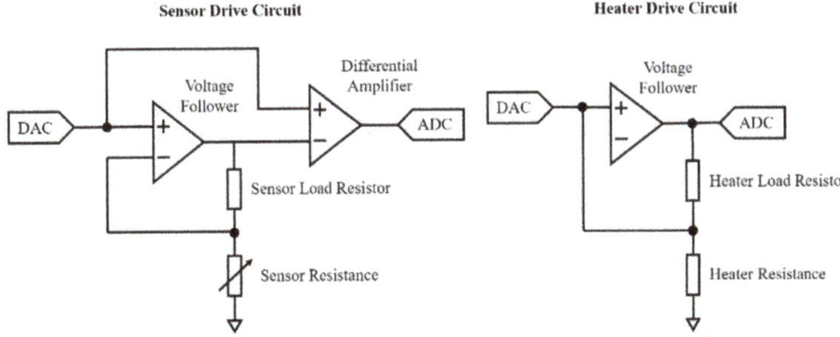


**Figure 6.** Simplified circuit diagram of sensor and heater drive circuits for analogue MOX sensors.

**Table 2.** Sensor and heater voltages and resistances for analogue VOC sensors.

| Sensor | Sensor Voltage | Sensor Resistance | Heater Voltage | Heater Resistance |
|---|---|---|---|---|
| MiCS-6814 * | 5.0 V | 180/45/650 kΩ | 2.4/1.7/2.2 V | 72/66/72 Ω |
| Dual Sensor | 5.0 V | 500 kΩ | 5.0 V | 50–500 kΩ |
| TGS-8100 | 3.0 V | 45 kΩ | 1.8 V | 110 Ω |
| TGS-2620 | 5.0 V | 20 kΩ | 5.0 V | 83 Ω |
| AS-MLV-P2 | 5.0 V | 500 kΩ | 3.0 V | 50–500 kΩ |

* MiCS-6814 has 3 VOC outputs: reducing gases, oxidising gases, $NH_3$.

### 2.4. Chemical Testing

The WOLF Breath E-nose was tested using chemical standards to calibrate the device and ensure that all sensors were functioning correctly. This is a standard approach used to calibrate gas sensor arrays [25]. For these tests, it is necessary to generate headspace gas from 3 chemicals; in this case, acetone, isopropanol and 1-propanol with 99% purity (Sigma-Aldrich, Dorset, UK). The chemicals were diluted using deionised water to concentrations of 0.1% acetone, 0.1% 1-propanol and 0.05% isopropanol. A set of 6 headspace vials were prepared for each chemical, with 1 mL in each. These were heated for 10 min at 40 ± 0.1 °C in a heater block (DB-2D Dri-Block, Techne/Cole-Parmer, Stone, UK). For testing, the vials were connected to the sensor chamber using PTFE tubing and the headspace gas was pulled into the sensor chamber using the miniature pump. The sequence of tested solutions was acetone, 1-propanol, isopropanol, [repeat] to mitigate against effects of sensor drift. After pulling the headspace gas into the chamber, the sample was disconnected to allow ambient air into the chamber. A period of 5–10 min was sufficient for sensor responses to return to baseline levels between each test. Principal component analysis (PCA) was conducted on the results. PCA is an unsupervised linear method, which reduces the dimensionality of the data by selecting a small number of linearly uncorrelated principal components (PC) that explain the majority of the variation in the data [34]. For PCA analysis, feature extraction was conducted using the difference model, whereby the maximum gas sensor response is subtracted from the stable baseline value, prior to the response [35].

In addition to these tests, the sensor array was evaluated using a gas rig to test the response to a single chemical, isobutylene, at different concentrations. A gas cylinder with a nominal concentration of 50 ppm was diluted with zero air from a zero-air generator (HPZA-7000-220, Parker, Warwick, UK) to concentrations of 2, 4, 6, 8, and 10 ppm using a custom mass flow-controlled gas mixing system. The total flow rate of the diluted gas was set to 300 mL/min and humidity was added using a water bubbler (around 60% RH, measured using the BME680 gas sensor). The test started with 1 hour of zero air, followed by 30-min of increasing concentration steps, with 30-min of zero air between each step. The experimental set-up for the gas rig testing is shown in Figure 7.

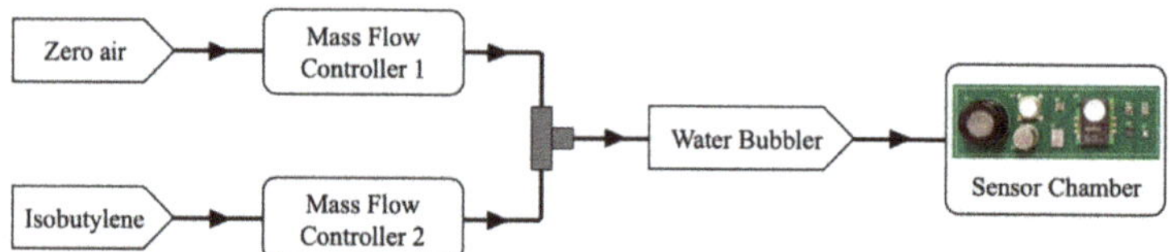


**Figure 7.** Experimental gas-rig set-up for testing isobutylene concentrations.

*2.5. Exhaled Breath Testing*

To simulate a control vs. disease group case-control study, a peppermint breath test was conducted. Ethical approval was obtained from local research ethics committee (BSREC reference: REGO-2018-2168). A total of 18 subjects were recruited for these tests. All subjects were healthy (self-reported), male, and between the ages of 21–30 (mean age of 25.4 years and standard deviation 1.8). Previous breath analysis studies have demonstrated that factors such as age, sex, weight, lifestyle and medication can influence breath composition [36]. To minimise the effects of these confounding factors, the recruited subjects were of the same sex and age (young adults, defined as ages 18–35 years).

Subjects provided breath samples pre- and post-consumption of a 200 mg peppermint oil capsule (2851512, Boots, Nottingham, UK). The intended application of these capsules is to support a healthy digestive system and aid the normal functioning of the digestive tract [37]. However, the consumption of peppermint oil also produces a well-defined, but temporary, change in the breath profile of an individual. As the ingested peppermint oil capsule dissolves, it releases volatile aroma compounds (e.g., menthol), which can be detected in breath for up to 10 hours [38]. Investigating the 'wash-out' profile of peppermint oil in breath has been proposed as a method of standardising breath analysis methods by benchmarking analysis techniques. Some work relating to this has already been published using GC-MS and PTR-MS technologies [38,39]. To the best of our knowledge, this is the first peppermint breath study conducted using E-nose technology. For this study, we used the standardised intervention of peppermint oil capsules to change the breath composition of the 'peppermint' breath samples to simulate a 'disease' group. The menthol intensity has been observed to peak after 30–45 min [39]. The sampling interval between 'baseline' and 'peppermint' samples was around 45-min for the conducted experiments.

## 3. Results

*3.1. Chemical Testing*

The PCA results are shown in Figure 8 (left) as a score plot. Each sample is presented by a single point, which demonstrates the relations (similarity) between all samples. Points that lie close to each other have similar properties while points that are further away have different properties [40].

This result demonstrates good separation between the three chemicals with no overlap. The principal components PC1 and PC2 account for over 90% of the variance, which is considered a robust result [41]. This indicates that the WOLF Breath E-nose has some selectivity at distinguishing between individual chemicals.

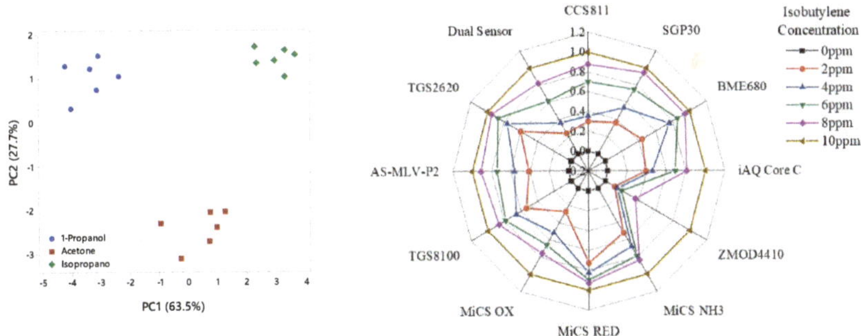

**Figure 8.** Testing of chemical standards; (**left**) PCA results; (**right**) Isobutylene concentration steps radar plot of normalised features.

The results from the isobutylene concentration gas rig tests are presented as a radar plot of normalised features in Figure 8 (right). The radar plot demonstrates that the sensors consistently responded to the concentration steps. Examples of the gas sensor responses from the SGP30 and TGS-2620 sensors are shown in Figure 9.

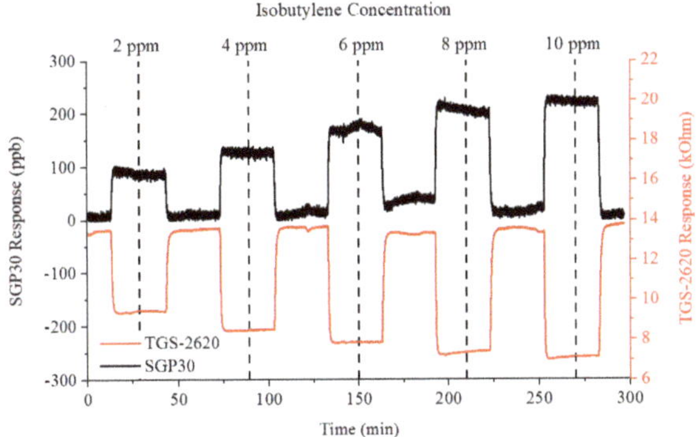

**Figure 9.** SGP30 and TGS-2620 gas sensor responses to increasing concentrations of isobutylene.

## 3.2. Volunteer Testing

Prior to conducting the peppermint breath tests, the device was turned on and left to stabilise in ambient air for over 1 h. A minimum start-up time of 10–15 min is necessary to allow the heater relay to heat the sampling tube to body temperature. Furthermore, the gas sensors need some time to reach operating temperatures and stable baseline output readings. The warm-up and baseline stabilisation output response is shown in Figure 10.

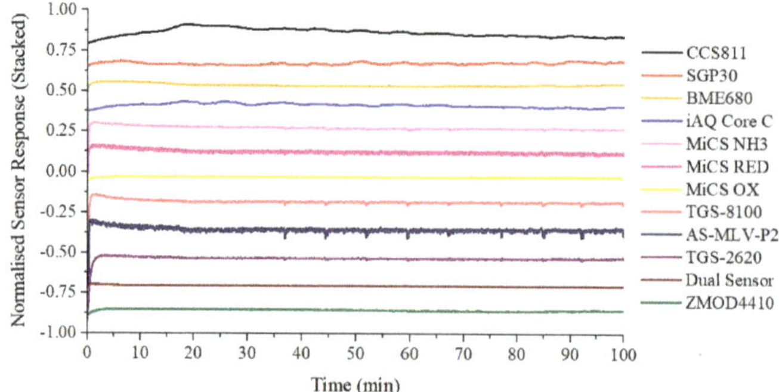

**Figure 10.** Warm-up and baseline stabilisation output response.

For our volunteer experiments, subjects did not need to exhale until their lungs were as empty as possible. Instead, only four seconds of exhaled breath were required. Subjects were asked to inhale for four seconds, and then exhale normally into the device for around four seconds. This procedure can improve the reproducibility of sampling, as shown in Figure 11. These readings are from the sensor module, embedded in the back-end connector of the sampling tube. Figure 11 demonstrates that using this standardised sampling procedure produces very consistent and reproducible outputs for commonly used sampling parameters, such as $CO_2$ and humidity. This procedure should therefore minimise sampling-related variability in exhaled breath composition. A typical sensor array output response to exhaled breath, from the same subject, is shown in Figure 12.

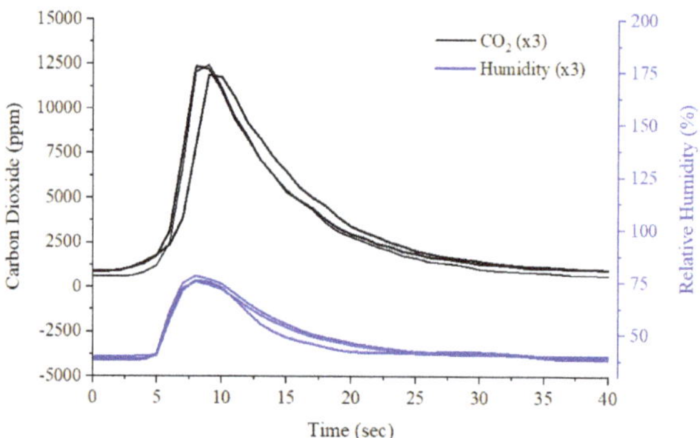

**Figure 11.** Breath sampling reproducibility using carbon dioxide and humidity sensor readings.

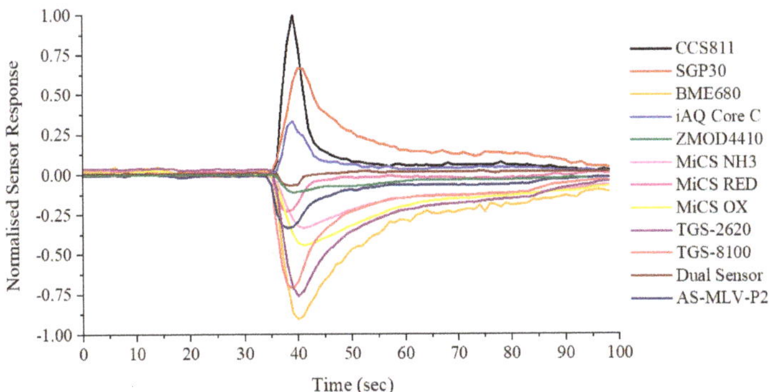

**Figure 12.** Typical output response to exhaled breath.

Most applications of exhaled breath research aim to discriminate between disease states. These applications require classification analysis to distinguish between groups and aim to build and train a model that can be used as a diagnostic tool. In our previous work, a standard classification analysis pipeline has been developed and applied to urinary and breath VOC studies using E-nose technology to investigate diabetes and inflammatory bowel disease (IBD) [42,43]. The same analysis pipeline was applied to this simulated case-control study between baseline (pre-consumption) and peppermint (post-consumption) exhaled breath samples. Classification results are expressed as a receiver operating characteristic (ROC) curve, as shown in Figure 13 (left). In ROC curves, the true positive rate (sensitivity) is plotted as a function of the false positive rate (1-specificity) [44]. The area under the ROC curve (AUC) is a measure of how well parameters can distinguish between groups [45]. An area of 1.0 represents a perfect test whereas an area of around 0.5 has no discriminatory power. Good separation would result in an AUC of around 0.7 or higher [46]. The classification results were AUC: 0.81, sensitivity: 0.83 (0.59–0.96), specificity: 0.72 (0.47–0.90), $p$-value: <0.001. These results were achieved using the Support Vector Machine (SVM) classifier. This classifier has been used in E-nose breath studies investigating cancer [47,48], pneumonia [49], and kidney disease [50]. When classification is performed, a $p$-value can be calculated to infer whether the null hypothesis is true, i.e. no differences exist between groups. Traditionally, a $p$-value smaller than 0.05 is used as an appropriate threshold to reject the null hypothesis [5]. A radar plot of normalised features is shown in Figure 13 (right). This demonstrates consistent differences in the sensor response between baseline and peppermint samples.

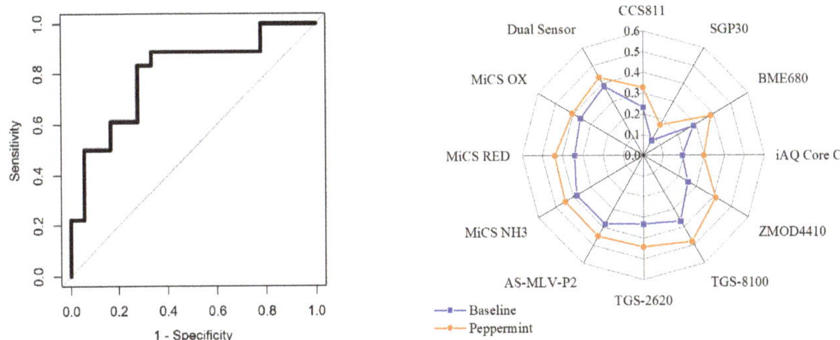

**Figure 13.** Baseline vs. peppermint exhaled breath samples; (**left**) ROC curve; (**right**) Peppermint breath test radar plot of normalised features.

## 4. Discussion

Testing of chemical standards, to discriminate between acetone, isopropanol and 1-propanol, and concentration steps of isobutylene, were conducted to evaluate the functionality of the developed unit. The results indicate that the WOLF Breath E-nose shows selectivity to distinguish between these chemicals. Different concentrations of isobutylene were associated with changes in the sensor array response. The concentration steps for the gas rig testing was 2 ppm. Since many exhaled VOCs are observed in low ppb range, further testing is required to determine the sensitivity of the gas sensors at lower concentrations. It should however be reiterated that the E-nose approach relies mainly on the cross-sensitivity of the gas sensors to interpret a complex sensor response pattern, as opposed to the individual sensor sensitivities to specific compounds.

The repeatability testing demonstrates that the sampling procedure can produce very consistent outputs. These readings can be used to monitor the quality and consistency of collected samples. For example, if multiple breath samples are to be collected from the same individual, the $CO_2$ output characteristics should be similar for each sample. If outliers or incorrect breathing (e.g., forced exhalation) is identified, the sample should be excluded or repeated. Volunteer testing of the system indicates that the unit is simple to use, for both the subject and operator, and comfortable for the subject. This sampling approach is non-restrictive and likely to be suitable for sampling vulnerable subjects, such as children or the elderly. The results from the peppermint study indicate that the WOLF Breath E-nose is capable of detecting the subtle change in breath composition that was induced by the consumption of peppermint oil. Restricting recruitment to young adult male subjects provided a robust experimental design. Future tests will include different age ranges and sexes, to reflect a more realistic sample group. Limited dietary control was enforced during the collection of exhaled breath samples. Strict dietary protocols and increasing the number of volunteers is likely to further improve classification results.

There are some limitations associated with the developed unit. A drawback of the MOX-based gas sensor unit is that there is an inherent variability between sensors, as a result of manufacturing processes [51]. To minimise these effects, it may be necessary to screen a large number of the same sensors to choose those with similar characteristics (e.g., sensor resistance at room temperature). While this increases development costs per unit, it is likely to be an effective method to reduce sensor variability between units. In addition to screening, the circuit design can be utilised to reduce possible problems of variability. For example, the heater drive circuit allows the real heater resistance to be calculated, which varies for different sensors of the same model. The required heater supply voltage can then be set so that the dissipated power is always the same.

If the proposed unit is intended to be used as a personalised breath analysis device, it will be necessary to be able to compare the results to other users. A calibration protocol needs to be developed to calibrate these devices regularly to evaluate whether the inter-variability between devices is acceptable for comparing breath profiles from different individuals. Such a protocol could involve benchmarking responses to a mixture of breath-related chemical standards. We have not yet evaluated how many times or how regularly the sensor array needs to be calibrated. Long-term stability testing of the system needs to be conducted in order to do this (e.g., tracking sensor drift over weeks or months).

Another limitation of MOX gas sensors is that the sensor response is affected by high levels of humidity. A strong correlation between changes in relative humidity and MOX sensor responses has been demonstrated [52]. However, since this E-nose was designed for applications in breath research, samples will inevitably have relatively high humidity levels. This was taken into account during gas rig testing by adding humidity using a water bubbler. Moreover, the effects of humidity will be relatively constant during the analysis of exhaled breath, since all samples will be associated with similar levels of high humidity.

As briefly discussed previously, sensor drift is also a critical factor for long-term monitoring applications of the E-nose. The experiments presented in this paper were conducted in a short period of a few days, where the external conditions of temperature and humidity were relatively constant.

These may be considered overoptimistic testing conditions, since practical long-term testing of the device would involve changes in background conditions (temperature, humidity, and composition of ambient air) [53]. Implementation of the duty cycle mode for analogue MOX sensors is likely to improve long-term stability of the unit; however, this needs to be further investigated. The electronic circuit design also allows for further development of modulating approaches, for both the sensor and heater drives.

## 5. Conclusions

In this paper, we have shown the development and testing of a compact, IoT-enabled, portable E-nose for breath analysis, which deploys an array of 10 commercial MEMS MOX gas sensors. The functionality of the device was demonstrated with the testing of chemical standards (discriminating between acetone, isopropanol and 1-propanol) and a gas rig testing of isobutylene from 2–10 ppm, in 2 ppm steps. PCA results indicate that the developed unit has some selectivity at distinguishing between individual chemicals. Gas rig testing results demonstrate that the sensor array consistently responded to the concentration steps. To further evaluate the developed system, the exhaled breath of 18 young male subjects was analysed before and after consuming a peppermint oil capsule. Classification analysis of exhaled breath samples demonstrates that we can distinguish between pre- and post-consumption of peppermint capsules with AUC: 0.81, sensitivity: 0.83 (0.59–0.96), specificity: 0.72 (0.47–0.90), $p$-value: <0.001. These results suggest that the unit can separate subject groups based on subtle changes in exhaled breath composition. It is our intention to deploy the unit in a UK hospital in an upcoming breath research study.

**Author Contributions:** J.A.C. conceptualised the project. J.A.C. and A.T. designed and configured the device. Samples were collected by A.T. Gas testing was completed by S.K.A. and A.T. Data analysis was conducted by A.W. and A.T. Original draft preparation, review and editing of the manuscript were completed by A.T. and J.A.C. All authors have read and agreed to the published version of the manuscript.

**Funding:** This research received no external funding. The PhD of A.W. is supported by Lembaga Pengelola Dana Pendidikan (LPDP), Ministry of Finance, Republic of Indonesia.

**Acknowledgments:** We would like to thank our volunteer subjects for participating in the peppermint study.

**Conflicts of Interest:** The authors declare no conflict of interest.

## References

1. Phillips, M. Breath Tests in Medicine. *Sci. Am.* **1992**, *267*, 74–79. [CrossRef] [PubMed]
2. Xu, M.; Tang, Z.; Duan, Y.; Liu, Y. GC-Based Techniques for Breath Analysis: Current Status, Challenges, and Prospects. *Crit. Rev. Anal. Chem.* **2016**, *46*, 291–304. [CrossRef] [PubMed]
3. Besa, V.; Teschler, H.; Kurth, I.; Khan, A.M.; Zarogoulidis, P.; Baumbach, J.I.; Sommerwerck, U.; Freitag, L.; Darwiche, K. Exhaled volatile organic compounds discriminate patients with chronic obstructive pulmonary disease from healthy subjects. *Int. J. Chronic Obstruct. Pulm. Dis.* **2015**, *10*, 399–406.
4. Altomare, D.F.; Di Lena, M.; Procelli, F.; Trizio, L.; Travaglio, E.; Tutino, M.; Dragonieri, S.; Memeo, V.; de Gennaro, G. Exhaled volatile organic compounds identify patients with colorectal cancer. *Br. J. Surg.* **2012**, *100*, 144–150. [CrossRef] [PubMed]
5. Pereira, J.; Porto-Figueira, P.; Cavaco, C.; Taunk, K.; Rapole, S.; Dhakne, R.; Nagarajaram, H.; Camara, J.S. Breath analysis as a potential and non-invasive frontier in disease diagnosis: An overview. *Metabolites* **2015**, *5*, 3–55. [CrossRef]
6. Miekisch, W.; Schubert, J.K. From highly sophisticated analytical techniques to life-saving diagnostics: Technical developments in breath analysis. *TrAC Trends Anal. Chem.* **2006**, *25*, 665–673. [CrossRef]
7. Bajtarevic, A.; Ager, C.; Pienz, M.; Klieber, M.; Schwarz, K.; Ligor, M.; Ligor, T.; Filipiak, W.; Denz, H.; Fiegl, M.; et al. Noninvasive detection of lung cancer by analysis of exhaled breath. *BMC Cancer* **2009**, *9*, 348. [CrossRef]

8.    Beauchamp, J.; Kirsch, F.; Buettner, A. Real-time breath gas analysis for pharmacokinetics: Monitoring exhaled breath by on-line proton-transfer-reaction mass spectrometry after ingestion of eucalyptol-containing capsules. *J. Breath Res.* **2010**, *4*, 026006. [CrossRef]

9.    Hunter, G.W.; Dweik, R.A. Applied breath analysis: An overview of the challenges and opportunities in developing and testing sensor technology for human health monitoring in aerospace and clinical applications. *J. Breath Res.* **2008**, *2*, 037020. [CrossRef]

10.   Dang, L.M.; Piran, M.J.; Han, D.; Min, K.; Moon, H. A Survey on Internet of Things and Cloud Computing for Healthcare. *Electronics* **2019**, *8*, 768. [CrossRef]

11.   Shurmer, H.V.; Gardner, J.W. Odour discrimination with an electronic nose. *Sens. Actuators B Chem.* **1992**, *8*, 1–11. [CrossRef]

12.   Kumar, P.; Deep, A.; Kim, K.-H.; Brown, R.J.C. Coordination polymers: Opportunities and challenges for monitoring volatile organic compounds. *Prog. Polym. Sci.* **2015**, *45*, 102–118. [CrossRef]

13.   Park, S.Y.; Kim, Y.; Kim, T.; Eom, T.H.; Kim, S.Y.; Jang, H.W. Chemoresistive materials for electronic nose: Progress, perspectives, and challenges. *Informat* **2019**, *1*, 289–316. [CrossRef]

14.   Wilson, D.A. Applications of Electronic-Nose Technologies for Noninvasive Early Detection of Plant, Animal and Human Diseases. *Chemosensors* **2018**, *6*, 45. [CrossRef]

15.   Wilson, A.D. Advances in electronic-nose technologies for the detection of volatile biomarker metabolites in the human breath. *Metabolites* **2015**, *5*, 140–163. [CrossRef] [PubMed]

16.   Saidi, T.; Zaim, O.; Moufid, M.; El Bari, N.; Ionescu, R.; Bouchikhi, B. Exhaled breath analysis using electronic nose and gas chromatography–mass spectrometry for non-invasive diagnosis of chronic kidney disease, diabetes mellitus and healthy subjects. *Sens. Actuators B Chem.* **2018**, *257*, 178–188. [CrossRef]

17.   Bruins, M.; Rahim, Z.; Bos, A.; van de Sande, W.W.J.; Endtz, H.P.; van Belkum, A. Diagnosis of active tuberculosis by e-nose analysis of exhaled air. *Tuberculosis* **2013**, *93*, 232–238. [CrossRef]

18.   Schnabel, R.M.; Boumans, M.L.L.; Smolinska, A.; Stobberingh, E.E.; Kaufmann, R.; Roekaerts, P.M.H.J.; Bergmans, D.C.J.J. Electronic nose analysis of exhaled breath to diagnose ventilator-associated pneumonia. *Respir. Med.* **2015**, *109*, 1454–1459. [CrossRef]

19.   Sánchez, C.; Santos, J.; Lozano, J. Use of Electronic Noses for Diagnosis of Digestive and Respiratory Diseases through the Breath. *Biosensors* **2019**, *9*, 35. [CrossRef]

20.   Kononov, A.; Korotetsky, B.; Jahatspanian, I.; Gubal, A.; Vasiliev, A.; Arsenjev, A.; Nefedov, A.; Barchuk, A.; Gorbunov, I.; Kozyrev, K. Online breath analysis using metal oxide semiconductor sensors (electronic nose) for diagnosis of lung cancer. *J. Breath Res.* **2019**, *14*, 016004. [CrossRef]

21.   Behera, B.; Joshi, R.; Anil Vishnu, G.K.; Bhalerao, S.; Pandya, H.J. Electronic nose: A non-invasive technology for breath analysis of diabetes and lung cancer patients. *J. Breath Res.* **2019**, *13*, 024001. [CrossRef] [PubMed]

22.   Santos, J.P.; Aleixandre, M.; Arroyob, P.; Suarez, J.I.; Lozano, J. An Advanced Hand Held Electronic Nose for Ambient Air Applications. *Chem. Eng. Trans.* **2018**, *68*, 235–240.

23.   Lozano, J.; Melendez, F.; Arroyo, P.; Suarez, J.I.; Herrero, J.L.; Camona, P.; Fernandez, J.A. Towards the Miniaturization of Electronic Nose as Personal Measurement Systems. *Proceedings* **2019**, *14*, 30. [CrossRef]

24.   Jaeschke, C.; Gloeckler, J.; El Azizi, O.; Gonzalez, O.; Padilla, M.; Mitrovics, J.; Mizaikoff, B. An Innovative Modular eNose System Based on a Unique Combination of Analog and Digital Metal Oxide Sensors. *ACS Sens.* **2019**, *4*, 2277–2281. [CrossRef] [PubMed]

25.   Westenbrink, E.; Arasaradnam, R.P.; O'Connell, N.; Bailey, C.; Nwokolo, C.; Bardhan, K.D.; Covington, J.A. Development and application of a new electronic nose instrument for the detection of colorectal cancer. *Biosens. Bioelectron.* **2015**, *67*, 733–738. [CrossRef] [PubMed]

26.   Lawal, O.; Ahmed, W.M.; Nijsen, T.M.E.; Goodacre, R.; Fowler, S.J. Exhaled breath analysis: A review of 'breath-taking' methods for off-line analysis. *Metabolomics* **2017**, *13*, 110. [CrossRef] [PubMed]

27.   Lourenço, C.; Turner, C. Breath Analysis in Disease Diagnosis: Methodological Considerations and Applications. *Metabolites* **2014**, *4*, 465–498. [CrossRef]

28.   Danesh, E.; Dudeney, R.; Tsang, J.-H.; Blackman, C.; Covington, J.A.C.; Smith, P.; Saffell, J. A Multi-MOx Sensor Approach to Measure Oxidizing and Reducing Gases. *Proceedings* **2019**, *14*, 50. [CrossRef]

29.   Madrolle, S.; Grangeat, P.; Jutten, C. A Linear-Quadratic Model for the Quantification of a Mixture of Two Diluted Gases with a Single Metal Oxide Sensor. *Sensors* **2018**, *18*, 1785. [CrossRef]

30.   Burgués, J.; Marco, S. Low Power Operation of Temperature-Modulated Metal Oxide Semiconductor Gas Sensors. *Sensors* **2018**, *18*, 339. [CrossRef]

31. Abidin, M.Z.; Asmat, A.; Hamidon, M.N. Comparative Study of Drift Compensation Methods for Environmental Gas Sensors. *IOP Conf. Ser. Earth Environ. Sci.* **2018**, *117*, 012031. [CrossRef]

32. Liu, H.; Zhang, L.; Li, K.; Tan, O. Microhotplates for Metal Oxide Semiconductor Gas Sensor Applications—Towards the CMOS-MEMS Monolithic Approach. *Micromachines* **2018**, *9*, 557. [CrossRef] [PubMed]

33. Vergara, A.; Llobet, E.; Martinelli, E.; Di Natale, C.; D'Amico, A.; Correig, X. Feature extraction of metal oxide gas sensors using dynamic moments. *Sens. Actuators B Chem.* **2007**, *122*, 219–226. [CrossRef]

34. Wlodzimirow, K.A.; Abu-Hanna, A.; Schultz, M.J.; Maas, M.A.; Bos, L.D.; Sterk, P.J.; Knobel, H.H.; Soers, R.J.; Chamuleau, R.A. Exhaled breath analysis with electronic nose technology for detection of acute liver failure in rats. *Biosens. Bioelectron.* **2014**, *53*, 129–134. [CrossRef] [PubMed]

35. Yan, J.; Guo, X.; Duan, S.; Jia, P.; Wang, L.; Peng, C.; Zhang, S. Electronic Nose Feature Extraction Methods: A Review. *Sensors* **2015**, *15*, 27804–27831. [CrossRef]

36. Blanchet, L.; Smolinska, A.; Baranska, A.; Tigchelaar, E.; Swertz, M.; Zhernakova, A.; Dallinga, J.W.; Wijmenga, C.; van Schooten, F.J. Factors that influence the volatile organic compound content in human breath. *J. Breath Res.* **2017**, *11*, 16013. [CrossRef]

37. The Boots Company. Boots Peppermint Oil 200 mg. Available online: https://www.boots.com/boots-peppermint-oil-200mg-60-capsules-10115320 (accessed on 8 April 2019).

38. Heaney, L.M.; Ruszkiewicz, D.M.; Arthur, K.L.; Hadjithekli, A.; Aldcroft, C.; Lindley, M.R.; Thomas, C.P.; Turner, M.A.; Reynolds, J.C. Real-time monitoring of exhaled volatiles using atmospheric pressure chemical ionization on a compact mass spectrometer. *Bioanalysis* **2016**, *8*, 1325–1336. [CrossRef]

39. Maláskavá, M.; Henderson, B.; Chellayah, P.D.; Ruzsanyi, V.; Mochalski, P.; Cristescu, S.M.; Mayhew, C.A. Proton transfer reaction time-of-flight mass spectrometric measurements of volatile compounds contained in peppermint oil capsules of relevance to real-time pharmacokinetic breath studies. *J. Breath Res.* **2019**, *13*, 046009. [CrossRef]

40. Jolliffe, I.T.; Cadima, J. Principal component analysis: A review and recent developments. *Philos. Trans. R. Soc. A Math. Phys. Eng. Sci.* **2016**, *374*, 20150202. [CrossRef]

41. Peng, G.; Tisch, U.; Adams, O.; Hakim, M.; Shehada, N.; Broza, Y.Y.; Billan, S.; Abdah-Bortnyak, R.; Kuten, A.; Haick, H. Diagnosing lung cancer in exhaled breath using gold nanoparticles. *Nat. Nanotechnol.* **2009**, *4*, 669–673. [CrossRef]

42. Esfahani, S.; Wicaksono, A.; Mozdiak, E.; Arasaradnam, R.P.; Covington, J.A.C. Non-Invasive Diagnosis of Diabetes by Volatile Organic Compounds in Urine Using FAIMS and Fox4000 Electronic Nose. *Biosensors* **2018**, *8*, 121. [CrossRef] [PubMed]

43. Tiele, A.; Wicaksono, A.; Kansara, J.; Arasaradnam, R.P.; Covington, J.A.C. Breath Analysis Using eNose and Ion Mobility Technology to Diagnose Inflammatory Bowel Disease—A Pilot Study. *Biosensors* **2019**, *9*, 55. [CrossRef] [PubMed]

44. Mandrekar, J.N. Receiver Operating Characteristic Curve in Diagnostic Test Assessment. *J. Thorac. Oncol.* **2010**, *5*, 1315–1316. [CrossRef] [PubMed]

45. Molinaro, A.M. Diagnostic tests: How to estimate the positive predictive value. *Neuro-Oncol. Pract.* **2015**, *2*, 162–166. [CrossRef]

46. Pearce, J.; Ferrier, S. Evaluating the predictive performance of habitat models developed using logistic regression. *Ecol. Model.* **2000**, *133*, 225–245. [CrossRef]

47. Hakim, M.; Billan, S.; Tisch, U.; Peng, G.; Dvrokind, I.; Marom, O.; Abdah-Bortnyak, R.; Kuten, A.; Haick, H. Diagnosis of head-and-neck cancer from exhaled breath. *Br. J. Cancer* **2011**, *104*, 1649. [CrossRef]

48. Machado, R.F.; Laskowski, D.; Deffenderfer, O.; Burch, T.; Zheng, S.; Mazzone, P.J.; Mekhail, T.; Jennings, C.; Stoller, J.K.; Pyle, J.; et al. Detection of Lung Cancer by Sensor Array Analyses of Exhaled Breath. *Am. J. Respir. Crit. Care Med.* **2005**, *171*, 1286–1291. [CrossRef]

49. Hockstein, N.G.; Thaler, E.R.; Torigian, D.; Miller, W.T., Jr.; Deffenderfer, O.; Hanson, C.W. Diagnosis of pneumonia with an electronic nose: Correlation of vapor signature with chest computed tomography scan findings. *Laryngoscope* **2004**, *114*, 1701–1705. [CrossRef]

50. Marom, O.; Nakhoul, F.; Tisch, U.; Shiban, A.; Abassi, Z.; Haick, H. Gold nanoparticle sensors for detecting chronic kidney disease and disease progression. *Nanomedicine* **2012**, *7*, 639. [CrossRef]

51. Zhang, L.; Tian, F.-C.; Peng, X.-W.; Yin, X. A rapid discreteness correction scheme for reproducibility enhancement among a batch of MOS gas sensors. *Sens. Actuators A Phys.* **2014**, *205*, 170–176. [CrossRef]

52. Sohn, J.H.; Atzeni, M.; Zeller, L.; Pioggia, G. Characterisation of humidity dependence of a metal oxide semiconductor sensor array using partial least squares. *Sens. Actuators B Chem.* **2008**, *131*, 230–235. [CrossRef]
53. Barsan, N.; Koziej, D.; Weimar, U. Metal oxide-based gas sensor research: How to? *Sens. Actuators B Chem.* **2007**, *121*, 18–35. [CrossRef]

*Article*

# A Computationally Efficient Mean Sound Speed Estimation Method Based on an Evaluation of Focusing Quality for Medical Ultrasound Imaging

Jaejin Lee [1], Yangmo Yoo [1,2], Changhan Yoon [3,*] and Tai-kyong Song [1,*]

[1] Department of Electronic Engineering, Sogang University, Seoul 04107, Korea; jge0707@sogang.ac.kr (J.L.); ymyoo@sogang.ac.kr (Y.Y.)
[2] Interdisciplinary Program of Integrated Biotechnology, Sogang University, Seoul 04107, Korea
[3] Department of Biomedical Engineering, Inje University, Kimhae 50834, Korea
* Correspondence: cyoon@inje.ac.kr (C.Y.); tksong@sogang.ac.kr (T.-K.S.); Tel.: +82-55-320-3301 (C.Y.); +82-2-707-3007 (T.-K.S.)

Received: 4 October 2019; Accepted: 16 November 2019; Published: 18 November 2019

**Abstract:** Generally, ultrasound receive beamformers calculate the focusing time delays of fixed sound speeds in human tissue (e.g., 1540 m/s). However, phase distortions occur due to variations of sound speeds in soft tissues, resulting in degradation of image quality. Thus, an optimal estimation of sound speed is required in order to improve image quality. Implementation of real-time sound speed estimation is challenging due to high computational and hardware complexities. In this paper, an optimal sound speed estimation method with a low-cost hardware resource is presented. In the proposed method, the optimal mean sound speed is determined by measuring the amplitude variance of pre-beamformed radio-frequency (RF) data. The proposed method was evaluated with phantom and in vivo experiments, and implemented on Virtex-4 with Xilinx ISE 12.4 using VHDL. Experiment results indicate that the proposed method could estimate the mean optimal sound speed and enhance spatial resolution with a negligible increase in the hardware resource usage.

**Keywords:** phase aberration; receive beamforming; spatial resolution; sound speed estimation

---

## 1. Introduction

Digital dynamic receive beamforming has been adopted for improving spatial resolution and contrast-to-noise ratios (CNRs) in medical ultrasound imaging [1–3]. In digital receive beamforming, a constant sound speed in human tissue (e.g., 1540 m/s) is typically assumed when generating dynamic receive focusing phase delays. However, phase distortions are introduced due to the variations of sound speeds that occur in soft tissue, leading to defocusing and consequent degradations in image quality [4]. Moreover, the degradation of image quality can significantly reduce diagnostic capability in breast or obese patient imaging, since the sound speed in fatty tissues (e.g., 1450 m/s) is lower than assumed value (e.g., 1540 m/s) [5].

Various phase aberration correction methods have previously been proposed to compensate for the phase distortions [6–13]. Representatively, cross-correlation-based [6–9] and speckle-brightness-based [10] phase aberration correction methods have been proposed. Recently, methods combined with adaptive beamforming methods have been proposed [11–13] that can improve the quality of whole images, but the implementation of these methods is still challenging due to their high computational complexity [13,14]. In the nearest-neighbor cross-correlation (NNCC) method [6,12,13], the number of multiplications is expressed approximately as $N_{mult} = (N-1) \times K_{image} \times L_{image} \times M$, where $N$, $L_{image}$ and $K_{image}$ are the number of channels, scanlines and samples per a scanline in the whole image, respectively, and $M$ is the total number of samples that contribute to the cross-correlation function. For an abdominal image with depth of 160 mm, the number of multiplications is approximately 1.3 billion $\times M$ when $N = 128$, $L_{image} = 256$ and $K_{image} = 4k$; thus, the implementation of these methods in real time would be challenging.

Using the correct mean sound speed can be an alternative solution for minimizing the phase distortions and enhancing the image quality in medical ultrasound imaging. Various algorithms for estimating sound speed have been proposed [14–20], and beam-tracking-based sound speed estimation methods using two transducers has been proposed [15,16]. This method can provide accurate local sound speeds; however, its clinical application is limited since the method uses two transducers. A direct estimation method that estimates the sound speed based on best-fit of one-way geometric delay patterns from the pre-beamformed radio-frequency (RF) channel data has also been proposed [17,18]. However, the performance of this method is sensitive to the transducer array geometry (i.e., the transducer's position) [18].

More recently, region-of-interest (ROI)-based optimal mean sound speed estimation methods have been proposed as a viable solution [14,19,20]. In these methods, the optimal mean sound speed that can produce the best focusing performance in the ROI is estimated rather than the actual sound speed in specific tissue type. The optimal mean sound speed is estimated using a lateral spatial fast Fourier transform (FFT) magnitude of beamformed data as the focusing quality factor [19]. Phase variations at each pre-beamformed RF channel data and coherent factors have been proposed as focusing quality factors to estimate the mean sound speed [14,20]. These methods have shown that image quality in the ROI can be improved by using optimal the mean sound speed in beamforming.

In this paper, we present a hardware-efficient optimal mean sound speed estimation method in which the focusing quality factor is measured by computing the minimum average sum of the absolute difference (MASAD) of pre-beamformed RF channel data, and thus enhance the spatial resolutions in medical ultrasound imaging. The proposed method estimates the mean sound speed that can provide improved image quality in the ROI for real-time imaging rather than full phase aberration correction. The proposed method was evaluated with phantom and in vivo experiments, and was implemented on a Virtex-4 FPGA (Field Programmable Gate Array) chip (Xilinx Instrument, San Jose, CA) with Xilinx ISE 12.4 using VHDL.

## 2. Materials and Methods

### 2.1. Minimum Average Sum of Absolute Difference (MASAD)

In ultrasound imaging systems, the dynamic receive focusing delays are computed based on the geometry of an ultrasound transducer and receive focusing points to adjust phase differences at each channel. The dynamic receive focusing delay of the $n^{th}$ element at $(x_n, z_n)$, for a focal point at $(x, z)$, is calculated by

$$\tau_n = \frac{\sqrt{(x - x_n)^2 + (z - z_n)^2} + R}{c} \tag{1}$$

where $R$ is a distance between the transducer center and the focal point, and $c$ is the assumed sound speed in soft tissues (e.g., 1540 m/s) [21]. After applying receive focusing phase delay in Equation (1), all RF channel data are coherently aligned when the assumed fixed sound speed is equal to the actual sound speed in a propagation medium. However, when the sound speed of a medium is different from the assumed one, the phase distortion cannot be avoided even when the receive dynamic focusing is employed. Since the phase is directly related to the amplitude change of the RF data, these phase distortions cause the amplitude variations [7].

In the proposed method, the optimal mean sound speed is determined by computing the minimum average amplitude variance between all RF channel data after applying the receive dynamic focusing in the region of interest (ROI). Furthermore, we utilized the sum of the absolute difference in place of computing the variance to reduce the computational complexity. Thus, the cost function of sound speed estimation is defined by

$$c_{opt} = \underset{c=c+c_{incr}}{argmin}\left[\frac{1}{LK}\sum_{l=0}^{L-1}\sum_{k=0}^{K-1}\sum_{n=0}^{N-1}\left|X_n(l,k) - X_{mean}(l,k)\right| * w(n)\right] \tag{2}$$

where $N$, $L$ and $K$ are the number of channels, scanlines and focal points per a scanline in the ROI, respectively. $X_n(l,k)$ is the delayed RF data of the $n^{th}$ element for the $k^{th}$ focal point at the $l^{th}$ scanline in the ROI, $X_{mean}(l,k)$ is the mean value of the delayed RF data and $w(n)$ is the window function. The ROI can be set to around or beyond a transmit focusing point to minimize the effect from phase distortion in the near field. In the proposed estimation method, the pre-beamformed RF data in the ROI are first captured after freezing the image upon a user's request. With an initial sound speed, the receive focusing delays are calculated using Equation (1). The focusing delays are applied to the captured RF data, and the average sum of the absolute difference (ASAD) in Equation (2) is computed. As indicated in Equation (2), the ASAD values are iteratively computed for the RF data in the ROI while changing the sound speed, and the sound speed providing the minimum ASAD (MASAD) value is determined as an optimal mean sound speed. Then, the estimated optimal mean sound speed is applied to subsequent real-time ultrasound beamforming to achieve the enhanced image quality.

Figure 1a shows a block diagram of a conventional dynamic receive beamformer (DRB) based on fractional delay beamforming architecture [22] with the proposed MASAD block, which is shaded gray. Pre-beamformed RF channel data from analog-to-digital converter (ADC) were 12 bit, which is typically used in medical ultrasound imaging systems. The signal to quantization noise ratio (SQNR) of typical ultrasound RF data is 74 dB, which is defined by SQNR $= 6.02b + 1.76(dB)$, where $b$ is number of bits of ADC (i.e., 12 bit) [23]. The block diagram of the proposed MASAD is shown in Figure 1b. As shown in Figure 1b, the MASAD block can be implemented with $N+2$ register, $N+2$ adder, $N$ absolute and $N$ multiplier where $N$ is the number of channels in the receive beamformer. To implement the proposed MASAD, input registers (pre-beamformed RF data), window coefficients and output register were 12 bits (12 integral bits), 8 bits (1 signed bit and 7 fractional bits) and 28 bits (21 integral bits and 7 fractional bits), respectively. Since we calculated the ASAD values without any truncation, the error between floating point calculation by PC and fixed point calculation by FPGA was less than 0.03%. The additional hardware for the MASAD block is not be a significant burden to the ultrasound imaging systems.

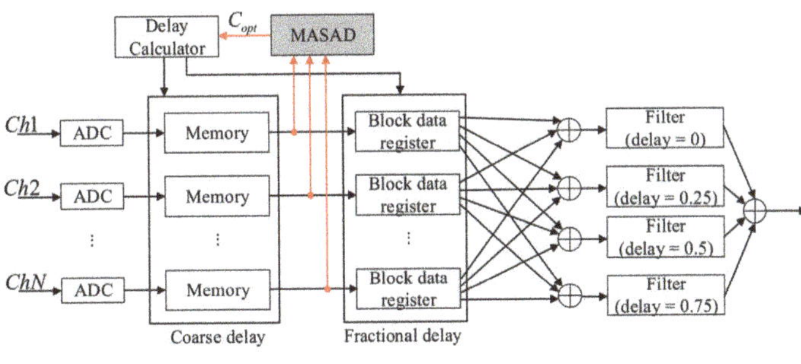


**(a)**

**Figure 1.** *Cont.*

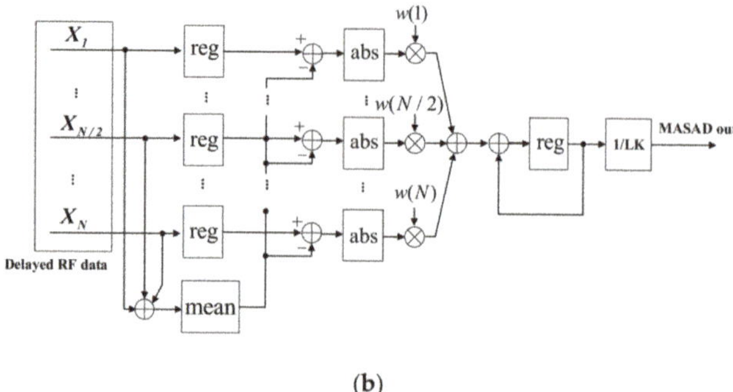

**(b)**

**Figure 1.** (**a**) Block diagram of the conventional dynamic receive beamformer (DRB) with the proposed MASAD and (**b**) the detailed block diagram of MASAD.

## 2.2. Experiment Setup and Evaluation Metrics

To evaluate the proposed MASAD method, an ultrasound research system (Vantage 128, Verasonics Inc., Redmond, WA, USA) was used to obtain pre-beamformed RF data using a 128-element convex array transducer (C5-2v, Ultrasonix Inc., British Columbia, Canada). The center frequency and sampling rate were 3.7 and 14.8 MHz, respectively. For the homogeneous medium experiment, a tissue-mimicking phantom (ATS 539, ATS Laboratories Inc., Bridgeport, CT, USA) with a sound speed of 1450 m/s (±1% error) was used. In the inhomogeneous medium experiment, a tissue-mimicking phantom (040 GSE, CIRS Inc, Norfolk, VA, USA), for which the sound speed was 1540 m/s (±1% error), was immersed in deionized water and pre-beamformed RF data were acquired. Note that the sound speed of deionized water at room temperature is 1480 m/s [24,25]. The in vivo data were acquired from the liver of a healthy volunteer. To compute the MASAD, the sound speed was changed from 1400 m/s ($c_{init}$) to 1600 m/s ($c_{max}$) in 10 m/s ($c_{incr}$) increments for each iteration process.

For quantitative evaluation, the lateral resolution was measured with the full width at half maximum (FWHM) for each of the two cases where the conventional sound speed (i.e., 1540 m/s) and the estimated one are employed. The contrast-to-noise ratio (CNR) values were computed for the cyst region by [26]

$$CNR = \frac{|\mu_c - \mu_s|}{\sqrt{(\sigma_c^2 + \sigma_s^2)}} \tag{3}$$

where $\mu_s$ and $\mu_c$ are the average intensities in the speckle and cyst regions, respectively, and $\sigma_s^2$ and $\sigma_c^2$ are the variances at each region.

The proposed MASAD sound speed estimation method was implemented on a FPGA (Virtex 4, Xilinx, San Jose, CA, USA) chip. In the experiments, the captured pre-beamformed RF data within a ROI were loaded in the custom-built FPGA platform [27], and then computed ASAD values were transferred to a PC to estimate the optimal sound speed. The hardware complexity was estimated by using Xilinx's ISE 12.4.

## 3. Results and Discussion

The results from homogeneous phantom experiments are shown in Figure 2. Figure 2a shows the B-mode image with a conventional sound speed (i.e., 1540 m/s); ROIs are indicated with white boxes to compute ASAD. The ROIs were selected around the transmit focusing point at 80 mm, and one ROI had a strong reflector while the other was a speckle region. Figure 2c shows the normalized ASAD values for each ROI. As can be seen, the estimated mean sound speed was 1460 m/s, which was within the fabrication error (i.e., 1450 m/s (±1% error)). The image with the estimated sound speed is shown in Figure 2b. Under the visual assessment, the improved spatial resolution can be readily identified when the estimated sound speed is used in dynamic receive beamforming. For quantitative comparison, the FWHM from the point target in the ROI was measured and showed 1.55 mm for the conventional sound speed and 1.05 mm for estimated one, indicating that the FWHM was improved by 33. 3% using the proposed method. The CNR of the cyst indicated with an arrow in Figure 2a was measured for each image, which were 2.83 in Figure 2a and 4.20 in Figure 2b; the CNR was improved by 48.4%. This result demonstrates that a higher CNR could be achieved using the estimated sound speed in beamformation.

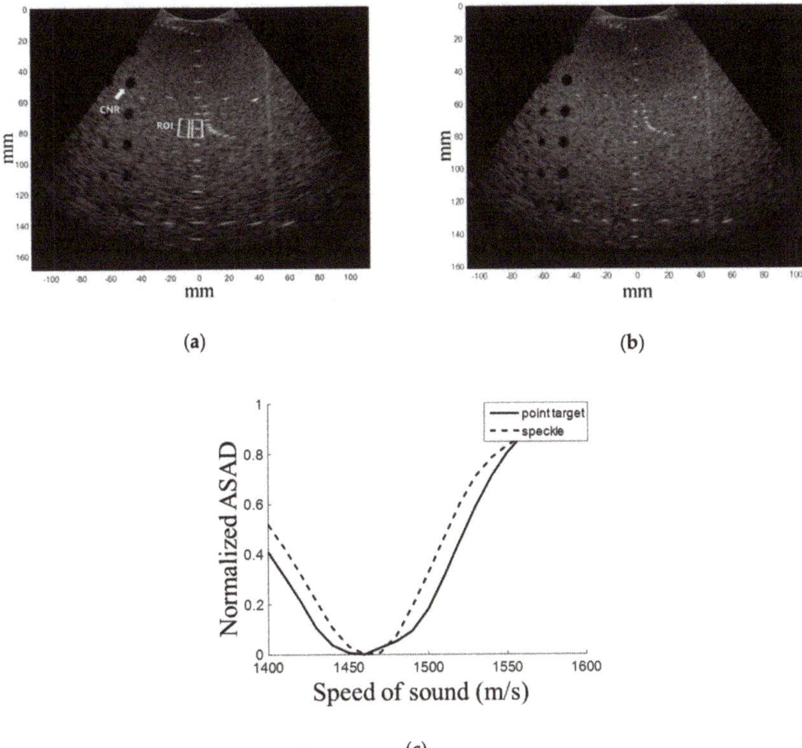


**Figure 2.** Experiment results of the homogeneous medium. B-mode image with a sound speed of (**a**) 1540 m/s, (**b**) 1460 m/s and (**c**) normalized ASAD values for the region of interest (ROI) of the point target (solid line) and speckle regions (dotted line).

The results of the inhomogeneous medium experiment are shown in Figure 3. In this experiment, the sound speeds were estimated for the ROIs with and without a strong reflector, which are indicated in Figure 3a. The ADAS values for each ROIs as a function of sound speed are plotted in Figure 3c,d. As shown in Figure 3c,d, the same sound speed (i.e., 1490 m/s) was estimated for ROI-1, -2 and 3 using the proposed MASAD, and 1500 m/s was estimated for ROI-4. The actual sound speeds for each ROI were calculated by $c = (d_w + d_p)c_w c_p / (c_p d_w + c_w d_p)$, where $d_w$ is the propagation distance in water (i.e., 50 mm), $d_p$ is the phantom (i.e., 10, 20, 30 and 40 mm) for each ROI; $c_w$ is the sound speed in water (i.e., 1480 m/s) and $c_p$ is the phantom (i.e., 1540 m/s); and the actual sound speeds for each ROI were 1490, 1497, 1502 and 1506 m/s, respectively. These discrepancies between the estimation results and actual sound speeds were within the range of fabrication error (i.e., ±1%). The B-mode image with the estimated sound speed (i.e, 1490 m/s) is shown in Figure 3b. As can be seen, the image with the estimated sound speed yielded a better quality image than the image obtained with a conventional sound speed (i.e., 1540 m/s). The FWHMs for ROI-1, -2, -3 and -4 with the estimated sound speed were 0.76, 0.69, 0.79 and 1.11 mm, respectively, while those with the conventional sound speed were 1.85, 1.34, 2.1 and 2.21 mm, respectively. The CNR values from the cyst indicated with a white arrow in Figure 3a were 1.89 and 2.53 for sound speeds of 1540 and 1490 m/s, respectively. The improvement of FWHM was 54.90% on average and CNR was improved by 33.9%.

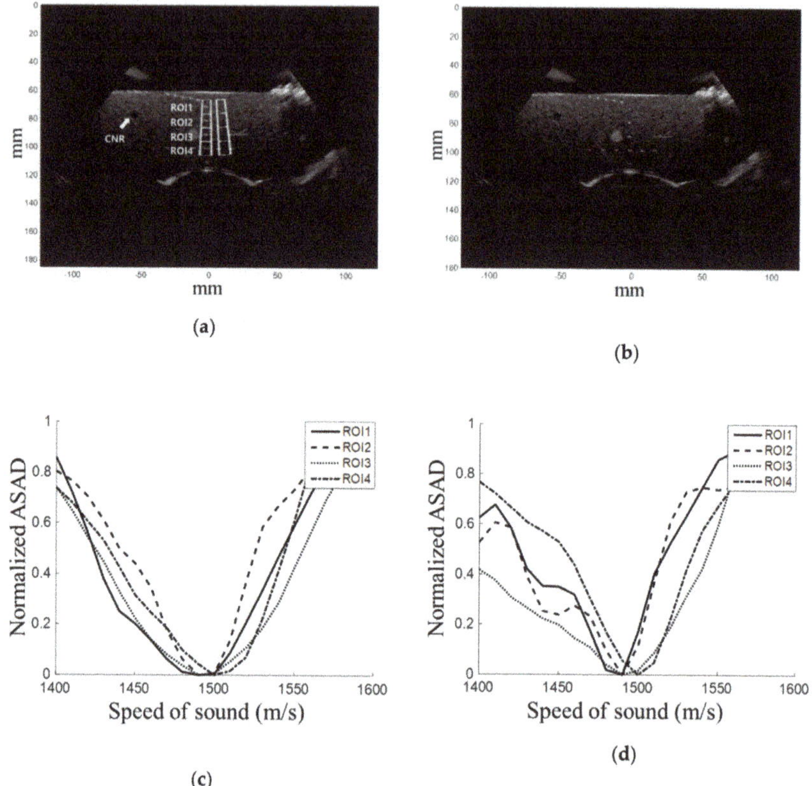


**Figure 3.** Experiment results of the inhomogeneous medium. B-mode image with a sound speed of (**a**) 1540 m/s and (**b**) 1490 m/s, and normalized ASAD values for ROI-1, -2, -3 and -4 for the (**c**) point target and (**d**) speckle regions.

Figure 4a,b shows the liver images constructed by using the conventional and estimated sound speeds, respectively. The ASAD as a function of sound speed is plotted in Figure 4c. As can be seen, the sound speed of 1580 m/s yielded MASAD, which is within the typical range of sound speed in human livers (i.e.,1550–1600 m/s) [28]. To clearly show the improvement of image quality, the profiles of the blood vessel wall in the white box are plotted in Figure 4d. The image with the estimated sound speed of 1580 m/s produced shaper edges than that with the conventional sound speed (i.e., 1540 m/s). The wall thickness with the conventional sound speed was 2.18 mm while it was 1.60 mm for the estimated one, indicating that the lateral resolution was improved by 26.6%. The CNR values inside the blood vessel improved 22.0%, and were 1.27 and 1.55 for the conventional and estimated sound speeds, respectively.

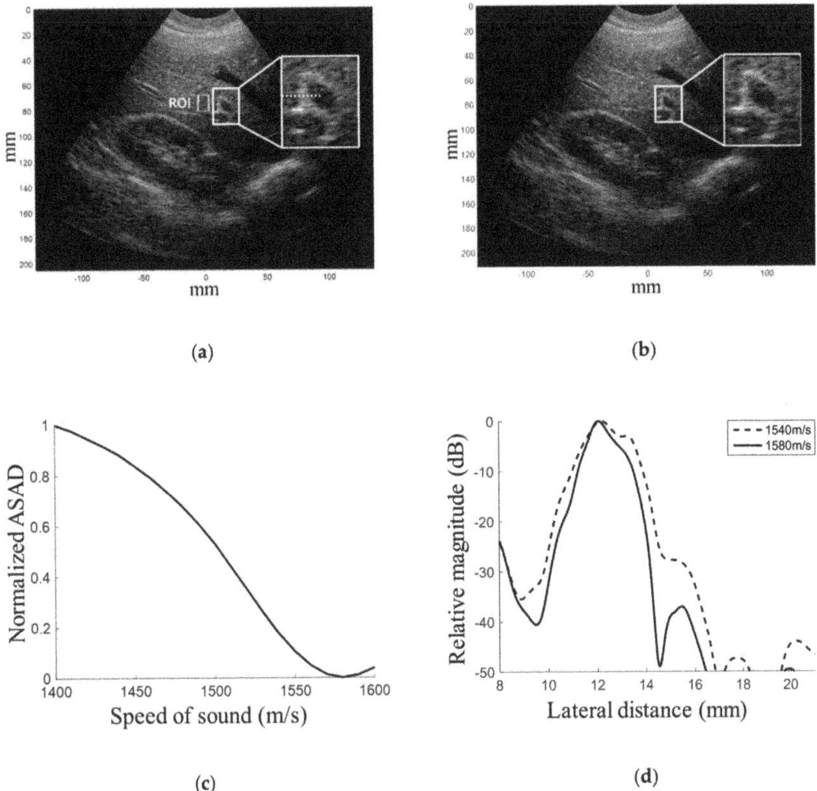

(a)

(b)

(c)

(d)

**Figure 4.** In vivo experiment results for the liver. B-mode image with sound speeds of (**a**) 1540 m/s and (**b**) 1580m/s, and (**c**,**d**) normalized ASAD values for the ROI. Cross-section of the vessel wall indicated by a dotted line in Figure 4a.

The time for estimating an optimal sound speed depends on the size of the ROI and is expressed as $t_{est} = N \times L \times K \times N_{iter} / f_{clk}$, where $N_{iter}$ is the number of iteration processes and $f_{clk}$ is the system clock used in the ultrasound imaging system. In the above experiments, we used $L = 10$, $K = 200$, $N = 128$, $N_{iter} = 20$ and $f_{clk} = 40$ Mhz, and the processing time was 0.128 s. Note that the estimation of the optimal sound speed was conducted to frozen images upon user request and subsequent beamforming was performed using the estimated sound speed; thus, the time for the processing would not affect the real-time operation.

The hardware complexity of the 32-channel dynamic receive beamformer with the MASAD block was estimated by utilizing the Xilinx ISE 12.4. The hardware utilization of the conventional DRB and the proposed DRB with the MASAD block are summarized in Table 1. As listed in Table 1, the hardware utilization of the proposed DRB with the MASAD slightly increased over the conventional DRB; 0.8% and 0.7% in slice LUTs and slice flip flops, respectively. These results indicate that the proposed method is capable of substantially improving the spatial resolution in medical ultrasound imaging and can be implemented with nearly a negligible increase in hardware complexity.

**Table 1.** Hardware resource of conventional and the proposed beamformers.

| Hardware Resource | Conventional DRB | Proposed DRB with MASAD |
|---|---|---|
| Slices | 24,659 (36%) | 24,722 (36%) |
| Flip Flops | 21,641 (16%) | 21,736 (16%) |
| Input LUTs | 28,994 (21%) | 29,083 (21%) |
| FIFO16/RAMB16s | 144 (50%) | 144 (50%) |

## 4. Conclusions

In this paper, a hardware-efficient mean sound speed method based on the minimum average sum of the absolute difference was presented for enhancing the spatial resolutions of medical ultrasound imaging. From the phantom and in vivo experiments, FWHM and CNR improved by an average of 46.41% and 134.77%, respectively. The proposed method demonstrated that it can improve spatial resolution with a negligible increase in hardware complexity. We believe that the proposed method would be a viable solution for estimating optimal sound speeds and could provide improved image quality. Further experiments in various clinical environments should be followed to validate the performance of proposed method.

**Author Contributions:** Conceptualization, supervision, and funding Acquisition, C.Y. and T.-K.S.; methodology, J.L. and C.Y.; validation, J.L. Y.Y. and C.Y.; writing—original draft preparation, J.L. and C.Y.; writing—review and editing, C.Y., Y.Y. and T.-K.S.

**Funding:** This work was supported in part by a National Research Foundation of Korea (NRF) grant funded by the Korea government (MSIP; Ministry of Science, ICT & Future Planning) (No. 2019R1A2C1089813) and the R&D program of MOTIE/KEIT (10076675, Development of MR Based High Intensity Focused Ultrasound Systems for Brain and Urinogenital Diseases).

**Conflicts of Interest:** The authors declare no conflict of interest.

## References

1. Song, T.; Park, S. A new digital phased array system for dynamic focusing and steering with reduced sampling rate. *Ultrason. Imaging* **1990**, *12*, 1–16. [CrossRef]
2. Steinberg, B.D. Digital beamforming in ultrasound. *IEEE Trans. Ultrason. Ferroelectr. Freq. Control* **1992**, *39*, 716–721. [CrossRef] [PubMed]
3. Mucci, R. A Comparison of Efficient Beamforming Algorithms. *IEEE Trans. Ultrason. Ferroelectr. Freq. Control* **1984**, *32*, 548–558. [CrossRef]
4. Anderson, M.E.; McKeag, M.S.; Trahey, G.E. The impact of sound speed errors on medical ultrasound imaging. *J. Acoust. Soc. Am.* **2000**, *107*, 3540–3548. [CrossRef] [PubMed]
5. Inagaki, K.; Arai, S.; Namekawa, K.; Akiyama, I. Sound Velocity Estimation and Beamform Correction by Simultaneous Multimodality Imaging with Ultrasound and Magnetic Resonance. *Appl. Sci.* **2018**, *8*, 2133. [CrossRef]
6. Flax, S.W.; O'Donnell, M. Phase-aberration correction using signals from point reflectors and diffuse scatterers: Basic principles. *IEEE Trans. Ultrason. Ferroelectr. Freq. Control* **1998**, *35*, 758–767. [CrossRef] [PubMed]
7. Karaman, M.; Atalar, A.; Koymen, H.; O'Donnell, M. A phase aberration correction method for ultrasound imaging. *IEEE Trans. Ultrason. Ferroelectr. Freq. Control* **1993**, *40*, 275–282. [CrossRef] [PubMed]
8. Li, P.C.; O'Donnell, M. Phase aberration correction on two-dimensional conformal arrays. *IEEE Trans. Ultrason. Ferroelectr. Freq. Control* **1995**, *42*, 73–82.
9. Behar, V. Techniques for phase correction in coherent ultrasound imaging systems. *Ultrasonics* **2002**, *39*, 603–610. [CrossRef]
10. Nock, L.; Trahey, G.E.; Smith, S.W. Phase aberration correction in medical ultrasound using speckle brightness as a quality factor. *J. Acoust. Soc. Am.* **1989**, *85*, 1819–1833. [CrossRef]
11. Mozumi, M.; Hasegawa, H. Adaptive beamformer combined with phase coherence weighting applied to ultrafast ultrasound. *Appl. Sci.* **2018**, *8*, 204. [CrossRef]

12. Ziksari, M.S.; Asl, B.M. Combined phase screen aberration correction and minimum variance beamforming in medical ultrasound. *Ultrasonics* **2017**, *75*, 71–79. [CrossRef] [PubMed]

13. Shin, J.; Yen, J.T. Synergistic enhancements of ultrasound image contrast with a combination of phase aberration correction and dual apodization with cross-correlation. *IEEE Trans. Ultrason. Ferroelectr. Freq. Control* **2012**, *59*, 2089–2101. [PubMed]

14. Cho, M.H.; Kang, L.H.; Kim, J.S.; Lee, S.Y. An efficient sound speed estimation method to enhance image resolution in ultrasound imaging. *Ultrasonics* **2009**, *49*, 774–778. [CrossRef] [PubMed]

15. Ophir, J. Estimation of speed of ultrasound propagation in biological tissues: A beam-tracking method. *IEEE Trans. Ultrason. Ferroelectr. Freq. Control* **1986**, *33*, 359–368. [CrossRef] [PubMed]

16. Robinson, D.E.; Ophir, J.; Wilson, L.S.; Chen, C.F. Pulse-echo Ultrasound Speed Measurements: Progress and Prospects. *Ultrasound Med. Biol.* **1991**, *17*, 633–646. [CrossRef]

17. Krucker, J.F.; Fowlkes, J.B.; Carson, P.L. Sound speed estimation using automatic ultrasound image registration. *IEEE Trans. Ultrason. Ferroelectr. Freq. Control* **2004**, *51*, 1095–1106. [CrossRef]

18. Anderson, M.E.; Trahey, G.E. The direct estimation of sound speed using pulse–echo ultrasound. *J. Acoust. Soc. Am.* **1998**, *104*, 3099–3106. [CrossRef]

19. Napolitano, D.; Chou, C.H.; McLaughlin, G.; Ji, T.L.; Mo, L.; DeBusschere, D.; Steins, R. Sound speed correction in ultrasound imaging. *Ultrasonics* **2006**, *44*, e43–e46. [CrossRef]

20. Yoon, C.; Lee, Y.; Chang, J.; Song, T.; Yoo, Y. In vitro estimation of mean sound speed based on minimum average phase variance in medical ultrasound imaging. *Ultrasonics* **2011**, *51*, 795–802. [CrossRef]

21. Kim, P.; Kang, J. A Pseudo-Dynamic Delay Calculation Using Optimal Zone Segmentation for Ultra-Compact Ultrasound Imaging Systems. *Electronics* **2019**, *8*, 242. [CrossRef]

22. Cho, J.; Lee, J.; Song, J.; Kim, Y.; Song, T. A fractional filter-based beamformer architecture using postfiltering approach to minimize hardware complexity. *IEEE Trans. Ultrason. Ferroelectr. Freq. Control* **2007**, *54*, 1076–1079. [CrossRef]

23. Kozmin, K.; Johansson, J.; Delsing, J. Level-crossing ADC performance evaluation toward ultrasound application. *IEEE Trans. Circuits Syst. I Regul. Pap.* **2008**, *56*, 1708–1719. [CrossRef]

24. Del Grosso, V.A.; Mader, C.W. Speed of sound in pure water. *J. Acoust. Soc. Am.* **1972**, *52*, 1442–1446. [CrossRef]

25. Bilaniuk, N.; Wong, G.S. Speed of sound in pure water as a function of temperature. *J. Acoust. Soc. Am.* **1993**, *93*, 1609–1612. [CrossRef]

26. Üstüner, K.F.; Holley, G.L. Ultrasound Imaging System Performance Assessment—AAPM. Available online: http://www.aapm.org/meetings/03AM/pdf/9905-9858.pdf (accessed on 9 August 2003).

27. Kang, J.; Yoon, C.; Lee, J.; Kye, S.; Lee, Y.; Chang, J.; Kim, G.; Yoo, Y.; Song, T. A system-on-chip solution for point-of-care ultrasound imaging systems: Architecture and ASIC implementation. *IEEE Trans. Biomed. Circuits Syst.* **2015**, *10*, 412–423. [CrossRef]

28. Bamber, J.C.; Hill, C.R. Acoustic properties of normal and cancerous human liver—I. Dependence on pathological condition. *Ultrasound Med. Biol.* **1981**, *7*, 121–133. [CrossRef]

 *electronics*

Article

# Incremental Low Rank Noise Reduction for Robust Infrared Tracking of Body Temperature during Medical Imaging

Bardia Yousefi [1,†], Hossein Memarzadeh Sharifipour [1], Mana Eskandari [1], Clemente Ibarra-Castanedo [1], Denis Laurendeau [1], Raymond Watts [2], Matthieu Klein [3] and Xavier P. V. Maldague [1,*]

1   Computer Vision and System Laboratory (CVSL), Department of Electrical and Computer Engineering, Laval University, Quebec City, QC G1V 0A6, Canada; Bardia.Yousefi.1@ulaval.ca (B.Y.); hossein.memarzadehsharifipour.1@ulaval.ca (H.M.S.); mana.eskandari.1@ulaval.ca (M.E.); clemente.ibarra-castanedo@gel.ulaval.ca (C.I.-C.); Denis.Laurendeau@gel.ulaval.ca (D.L.)
2   RT Thermal Co., 7167 Elkhorn Drive, West Palm Beach, FL 33411, USA; rwatts@rtthermal.com
3   Visiooimage Inc., 2560, Rue Lapointe, Sainte-Foy, Quebec City, QC G1W 1A8, Canada; matthieu.klein@visiooimage.com
*   Correspondence: Xavier.Maldague@gel.ulaval.ca; Tel.: +1-418-656-2962
†   Current address: Department of Radiology, University of Pennsylvania, Philadelphia, PA 19104, USA.

Received: 21 September 2019; Accepted: 4 November 2019; Published: 7 November 2019

**Abstract:** Thermal imagery for monitoring of body temperature provides a powerful tool to decrease health risks (e.g., burning) for patients during medical imaging (e.g., magnetic resonance imaging). The presented approach discusses an experiment to simulate radiology conditions with infrared imaging along with an automatic thermal monitoring/tracking system. The thermal tracking system uses an incremental low-rank noise reduction applying incremental singular value decomposition (SVD) and applies color based clustering for initialization of the region of interest (ROI) boundary. Then a particle filter tracks the ROI(s) from the entire thermal stream (video sequence). The thermal database contains 15 subjects in two positions (i.e., sitting, and lying) in front of thermal camera. This dataset is created to verify the robustness of our method with respect to motion-artifacts and in presence of additive noise (2–20%—salt and pepper noise). The proposed approach was tested for the infrared images in the dataset and was able to successfully measure and track the ROI continuously (100% detecting and tracking the temperature of participants), and provided considerable robustness against noise (unchanged accuracy even in 20% additive noise), which shows promising performance.

**Keywords:** infrared and thermal image analysis; incremental low rank noise reduction; incremental singular value decomposition; segmentation; monitoring of body temperature; particle filter tracking

---

## 1. Introduction

Thermographic applications in medicine have been increasing over past years [1–3] and a great variety of research has been conducted in different fields of medicine such as pain diagnosing and treatment monitoring [4], breast cancer [5], psychology [6], dentistry [7], avian flu [8], and many other applications. The application of thermography focuses on medical prevention and monitoring particularly during radiology imaging, where external measurement tools could be the source of a risk for the health of patients ([9–13]). An infrared tracking tool is proposed with robustness against noise to tracking body temperature and is applicable to patients during medical imaging and can be used for different modalities in radiology (one good example is functional magnetic resonance imaging (fMRI) [14], or recording devices such as electroencephalogram (EEG) during fMRI [15] and MRI [16–18]).

Medical imaging provides vital information from patients and plays a big role in diagnosis and prognosis of diseases. Among all medical imaging modalities, magnetic imaging [19,20] avoids exposure to X-ray radiation [21] and involves no side-effects for the human body. However, there are some reported cases of burning or issues for some patients with implants (e.g., metal or breast implants [22,23]). Metallic materials, chips, foreign objects such as artificial joints, prosthetic devices, pacemakers, metallic bone plates, and surgical clips can considerably affect the MRI imaging. In addition, heart pacemakers, metal implants or metal clips in (or even around) the patient's eyes cannot be scanned by MRI due to the risk of metal objects movement in the magnetic field. This involves bullet fragments, artificial heart valves, metallic ear implants, insulin pumps and even chemotherapy patients. There are some cases of joint pain (hip joint) and tenderness, rib cage pain from fibrosing disease called Nephrogenic Systemic Fibrosis (NSF).

One of the most probable issue is related to the adjustment of radiology instruments that is not properly tuned for the patient. Electromagnetic exposure of the patient's body has some biological effects which can be categorized as thermal effects and non-thermal effects [24]. These effects are due to direct energy transformation to living tissues and is associated with the frequency of the field [25]. Applying heating to the human body has different effects, and mostly depends on its sensitivity [26,27]. For example, permanent cosmetics and tattoos on the skin with metallic pigments (i.e., iron oxide) may cause first or second degree burns on a patient's body [28–30]. To alleviate such burns, dosimetric parameters are commonly used as a safety standard, specially for the absorption rate (e.g., specific absorption rate (SAR)) [31]. The standard level of SAR is approximately equivalent to an increase in scanning temperature of 0.6 °C for 20–30 min time duration [32–34], which prevents hot spot occurrence [35,36]. There are very few systems for early detection or prevention of such burns problem. There is no possibility to insert a probe in the body since all metal is prohibited on the body during the scanning. However, these complications can never overshadow the significance of medical imaging as these instruments play vital role in diagnosis, and prognostic of diseases. Nevertheless, the issue of over-heating remains partially unsolved. The objective of this research is to determine and track the overheating spots using an infrared imaging system during radiology examination. There are some similar approaches involving the tracking of body temperature (e.g., [37]) using a particle filter to track the selected the region of interest (ROI) [38] or combined with a Kalman filter to mitigate the effect of noise in tracking and thermal measurements [39] in the human body. The proposed approach creates a thermal dataset from several volunteers in a lying position (similar to radiology imaging environments) in front of thermal camera. For every frame, a color based segmentation of the ROI was performed in the thermal image. Then, this region fed to a particle filter to track it throughout the stream. Noise is an important problem which aggravates the performance in such systems. Several research studies have been conducted to reduce or model noise in thermal imagery (e.g., [40–42]). The proposed approach can monitor and observe the body temperature from a distance (with initial calibration) which eliminates the possibility of direct contact with the patient while tracking the temperature. Moreover, it implements a noise reduction approach which renders our system more robust against environmental noise. The proposed system was challenged by additive noise to a thermal image dataset created by 15 participants and achieved a very promising performance. This paper starts with stating the problem. Then a brief review on burning and dermatological effects is described next. The possibility of a 3D reconstruction of the patients in visible imaging is discussed afterwards. Section 2 describes the methodology for the incremental low rank noise reduction and tracking system, and is followed by the presentation of experimental results (Section 3). We discuss the challenges and advantages of the approach in Section 4. Finally, we conclude and describe future work in Section 5.

*3D Analysis*

For analysis of the human body temperature, a 3D thermal model of the body can be constructed during MRI exposure (a visible 3D reconstruction is shown in Figure 1). By 3D analysis of the infrared image, the thermal 3D model of the body allows the overheating points to be detected. In this section, a brief review of 3D processing analysis is presented. The human body is a living organism which has changes in internal factors and external form and is constantly in motion. These variations involve many factors such as pose shifts, fluid distributions in the body, sway, respiration and occlusion which create many limitations and difficulties in the topology of the body. Scattering properties, skin pigmentation and especially the radiology gown are considered as limitations that create problems in accurate measurement. The use of infrared images helps to overcome the mentioned difficulties. However, infrared images have their limits and face their own barriers. An explanation of the indoor thermal radiation environment in the human body has been analyzed through 3D modeling of the human body in some poses and situations [43].

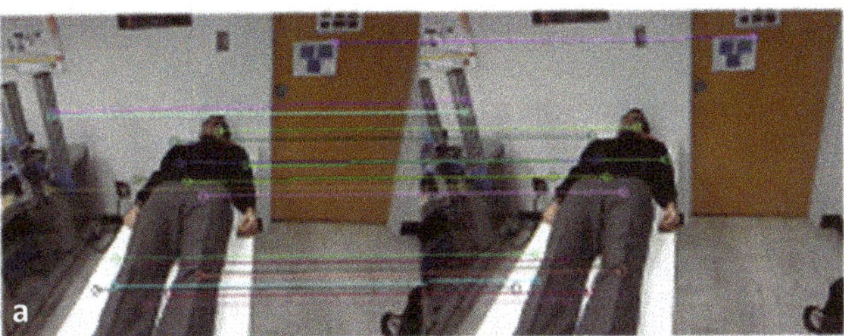

SURF descriptors in visible images for 3D construction

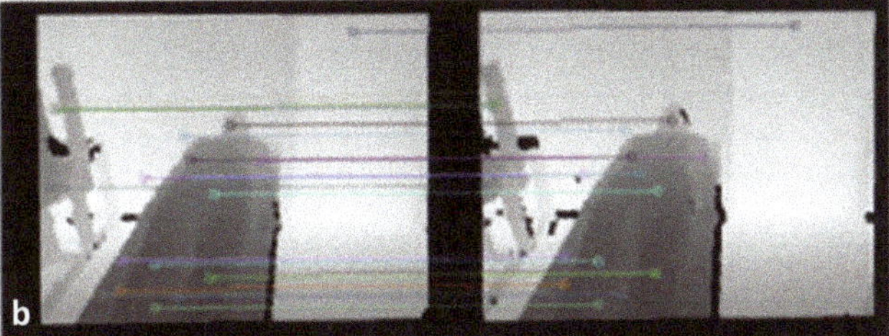

Similar points using SURF descriptors in depth images using Kinect camera

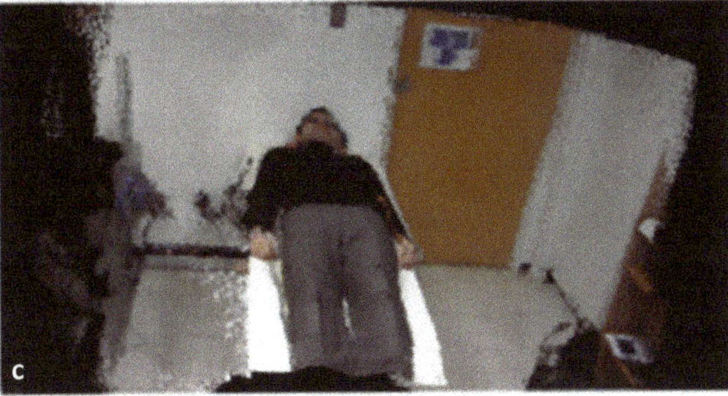

Point clouds (reconstructed 3D)

**Figure 1.** An example of making 3D model of a participant using a Kinect camera. (**a**) shows a visible image. (**b**,**c**) are depth images and reconstructed 3D images applying speeded up robust features (SURF), finding most similar keypoints, using iterative closest points (ICP), and point cloud library (PCL) for visualization, respectively.

A 3D X-ray microtomography image analysis of low density wood fiberboard has been conducted. It was used for the prediction and modeling of local densification with material behaviour under radiation exposure [44]. The study in 3D visualization of fiber is not limited to this investigation and has been studied in greater detail (for example [45]). There is another application of 3D reconstruction using thermal images for detection of infection in the skin of patients. An integration of 3D and high resolution far-infrared FIR thermal images of the body has been used for 3D structural binocular profilometer [46]. Research on jet fuels toxicity concerning skin damage measured by 900 MHz skin microscopy has been conducted. The authors applied three-dimensional spatial visualization that can show the skin structure and help draw conclusion regarding the toxicity [47].

Many medical image processing research studies have been conducted on 3D visualization of limbs and tissues under MRI scanning. The structural anatomy which has been considered by some 3D applications can be summarized as follows: vertebrae and Spin; Skull and brain (head); general, hip joint, femur, tibia, knee joint (limbs); perineum, entire (pelvis); heart, ribs, entire (thorax) [48]. Most research work on 3D (or even 2D) in biomedical image analysis adding the following areas: patient motion tracking, patient positioning (radiotherapy), arteriovenous malformation (radio-surgery), embolization, shunt angioplasty (vascular interventional radiology), neck, head, and spine, procedures which are invasive (neuroradiology interventional), replacement of hip and knee, total arthroplasty hip (orthopedic surgery), knee kinematics (kinematic study) with benefits for radiosurgery and radiotherapy applications [48]. A 3D building construction using Infrared thermography images and real images has been presented [49] which is not categorized in medical image analysis but the idea is valuable in terms of the potential applications.

Applying an oriented scheme cubic for co-registration of the practical geometry of medical parts of the human body using points and 3-plane has been presented based on localization properties and fusion of MR, computed tomography (CT) and positron emission tomography (PET) images for creating a 3D model [50]. The model includes images fused across modalities of PET+MR, CT+MR, PET+CT, MR+CT+PET and has been tested on patients for detectability of tumors with significant results. However, no thermal images or skin visualization was done. A very relevant research work regarding 3D surface thermal image construction has been conducted for energy auditing which gives a 3D surface temperature model [51]. This work is relevant however it was done for nonliving objects. Applying this technique to humans would make it a very useful system for relevant applications. Siewert et al. (2014) present a method for the analysis of body temperature in pig skin using thermal infrared images. This approach has used the averaging value of the temperature for both ROI (from two anatomical regions in IR images) to reduce noise [52]. One of the drawbacks of 3D reconstruction of the patient during imaging besides the difficulties of setting points as references, relates to reconstruction of 3D with thermal imaging. After camera calibration, a 3D object using two (or more) photos from different angles (limited angles, e.g., less that 90 degree) is reconstructed by following steps:

1.  Registration of depth to RGB (in order to align depth map with RGB image)
2.  Applying speeded up robust features speeded up robust features (SURF) [53] to both images to find keypoints;
3.  Comparing keypoints' descriptors to find most similar points
4.  Estimating the required rotation and translation matrices to register keypoints pairs using iterative closest points (ICP) [54]
5.  Performing the transformations and calculating the final 3D coordinates
6.  Accumulating the 3D coordinates and their colors in the final point cloud matrix
7.  Visualize the point cloud using point cloud library (PCL) [55].

These steps were implemented in two phases, the first phase was to generate a 3D model of an object using a left view (first image) and a right view (second image) of the object. Then, to reconstruct a more detailed 3D model using an image sequence captured by a Kinect camera while it was moving in front of the patient with limited angle. Figure 1 presents an example of making 3D model of a patient (participant) using a Kinect camera with limited degree of freedom designed for installation in front of imaging system. The information of depth and visible was used to reconstructed 3D images applying speeded up robust features (SURF), finding most similar key-points, using iterative closest points (ICP), and point cloud library (PCL) for visualization.

## 2. Incremental Low Rank Robust Tracking

The necessity of employing the thermal infrared system to monitor thermal changes for medical applications was discussed in the previous section. Here, the proposed method for monitoring the thermal variations is presented (Figure 2). The process begins with lower rank noise reduction and then an automatic detection of the ROI.

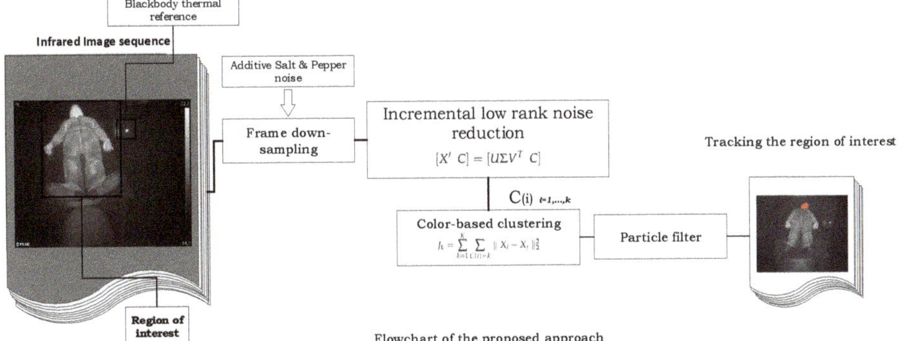


**Figure 2.** Flowchart which shows the proposed approach for the estimation of down-welling radiance among all of the possible points.

### 2.1. Low Rank Noise Reduction

Due to the nature of the process involving thermal cameras, the presence of noise in the thermal images seems inevitable. The sensitivity of the acquisition also depends on the acquisition conditions which may decrease the Signal to Noise Ratio (SNR) in the system.

Let $X$ be an input matrix which has dimension $p \times q$ where $p$ is the number of frames and $q$ is the vector corresponding to each image ($q = image\ height(m) \times image\ width(n)$). The Singular Value Decomposition (SVD) [56] gives a bilinear factoring of input matrix $X$, ($X = U\Sigma V^T$) which can be shown to decrease the rank by r and ultimately reduce the noise level as the noise's eigenvectors correspond to the lower eigenvalues in $\Sigma$. The mentioned process can described as follows:

$$X'_{(p \times q)} \overset{SVD_r}{\rightarrow} U_{p\times r}\Sigma_{r\times r}V^T_{r\times q},\ r \le min(p,q) \tag{1}$$

For higher $r$ in low-rank representation, the data is represented in a more explanatory form due to noise reduction ($X'$).

### 2.2. Incremental SVD

Performing the low rank noise reduction by SVD Equation (1) for the purpose of video processing and possible online system enforces the SVD to function incrementally. Incremental SVD [57] can handle randomly missing and incomplete or uncertain data. It is potentially an efficient robust subspace projection. We already have the eigenvalue decomposition of $X$ ($U\Sigma V^T$) and for the additional matrix $C$ to the previous data ($X$), we have:

$$[X'\ C] = \left[U\Sigma V^T\ C\right] \tag{2}$$

$$\left[U(1-UU^T)C/M\right]\begin{bmatrix} \Sigma & Y \\ 0 & M \end{bmatrix}\begin{bmatrix} V & 0 \\ 0 & 1 \end{bmatrix}^T \tag{3}$$

$$[U\ \Psi]\begin{bmatrix} \Sigma & Y \\ 0 & M \end{bmatrix}\begin{bmatrix} V & 0 \\ 0 & 1 \end{bmatrix}^T \tag{4}$$

Equations (2) and (3) are representing the procedure of adding the additional matrix $C$ (as a new infrared image from the stream) added into previous decomposition ($X$ representing the components from the previous batch of infrared images). Where $\doteq (I - UU^T)C = C - UL, Y = U^TC$ and $M \doteq \Psi^T T$

where $\Psi$ is an orthogonal basis of $T$. Applying QR-decomposition (QR of an orthogonal matrix Q and an upper triangular matrix R), there will be $T \overset{QR}{\to} \Psi M$. In the Equation (3), the middle matrix is denoted as $Q$ and must be updated and diagonalized to lead to the answer for the whole process. In the case of adding a single vector $c = C$ the computation speed is increased by calculating the vector $\psi = \Psi = (c - UU^T c)/m$ and scalar $c^T c - 2Y^T Y + (UY)^T (UY) \to m$. First $m \in M, \|m\| \to \Sigma, m/\|m\| \to U$, and $V \to 1$ and is then updated by iteration for the above mentioned calculations with truncation [58]. In the proposed approach, the initial data matrix is constructed using initial frames and after noise reduction, the additional frame ($c$) is added to the current data ($X'$) to complete the process.

### 2.3. Clustering and Tracking

After noise reduction for the initialization of the particle filter, a color based K-means clustering [38,59] is used. Here, a brief review of the kernelled K-means [60] is presented. Let $S = \{X_1, \cdots, X_n\} \subset R^p$ be our observation set ($p$) and dissimilarities are calculated by $\|X_i - X_j\|_2^2$ for $X_i \in R^p$ and $K$ is the number of clusters. Clustering of the data $X$ is nothing more than a function $C$ assigning every observation $X_i$ to a set of $k \in \{1, \cdots, K\}$. The minimization of the following formula is required:

$$J_k = \sum_{k=1}^{K} \sum_{C(i)=k} \|X_i - X_j\|_2^2 \tag{5}$$

Let $X = 1/n \sum_{i=1}^{n} X_i$ and $C(i) = k$ means that $X_i$ is assigned to group $k$. A different observation here is closely related to a different color base $X$. The color based clustering determines the ROI and is labelled using the reference temperature provided by the blackbody during the experiment or in thermal calibration process (one time for all). Tracking these labels is done using a particle filter which is similar to the previously presented approaches [38]. The function updates the ROI and concentrates the particles within the video stream.

### 2.4. Particle Filter

Clustering analysis detects and groups hot-spots and then a particle filter tracks these hot-spots with temperature updates and thermal expansion during imaging occurs (experiment). The robustness of tracking algorithm plays an important role due to presence of thermal fluctuations and noise (which is suppressed by the propose algorithm). Particle filter tracking algorithm [61,62] is employed for tracking and adaptation of thermal variations ( [63–65]), and provides a reasonably robust system within thermal imaging stream. The following assumptions have been considered for the tracking algorithm:

- Thermal images are gray scale (0,255) corresponds to cold and hot representation;
- Thermal camera's field of view (FOV) always has ROI;
- The ROI's temperature is higher than the surrounding temperature;
- The ROI does not have a particular shape and is adjustable in the algorithm with the respect to thermal increases (elevating image intensity);
- The ROI updates during the experiment (simulating medical test) and temperature updates by an upward trend to find hot spots which are cause of the burning in patients.

Particle filter performs in time $t$ and approximates tracking the target recursively by a finite set of posterior distribution weighted samples. Particle filters simulate the class filters for approximating random variables recursively. Let $\alpha_t | Y_t = (y_1, y_2, \cdots, y_t)$ be the random variables and $\alpha_t^1, \cdots, \alpha_t^M$ are particles, which have discrete probability mass of $\pi_t^1, \cdots, \pi_t^M$. Points for variable approximation are shown by $f(\alpha_t | Y_t)$ and for $\pi_t^j$ are assumed to be equal to $1/M$, which is the preferred amount of $M$ for particles to approximate the density value of $\alpha_t | Y_t$. It is noticeable that particles are located in the ROI, which is previously defined, and incrementally updated throughout experiment (medical exam). The discrete support is used as true density and provides an approximation of density prediction using particle support and empirical prediction:

$$\hat{f}(\alpha_{t+1}|Y_t) = \sum_{j=1}^{M} \hat{f}(\alpha_{t+1}|\alpha_t^j)\pi_t^j \tag{6}$$

Mixture of echoes while the filtering proceed and density. This provides the following modification on previous equation:

$$\hat{f}(\alpha_{t+1}|Y_{t+1}) \propto \hat{f}(y_{t+1}|\alpha_{t+1}) \sum_{j=1}^{M} \hat{f}(\alpha_{t+1}|\alpha_t^j)\pi_t^j \tag{7}$$

The above-mentioned equation is an approximation of true density filtering. New particles are produced $\alpha_{t+1}^1, \cdots, \alpha_{t+1}^M$ with weights $\pi_{t+1}^1, \cdots, \pi_{t+1}^M$ and this iterates throughout thermal imaging stream, which includes online tracking problems and an estimation of one-step-ahead density $\hat{f}(y_{t+1}|Y_t)$. This is relevant to updates of the ROI and spreading of hot spots during radiology exam ( [61,66]).

## 3. Results

### 3.1. Experimental Setup and Thermal Image Database

The experiments were conducted at room temperature using an A65 FLIR camera (Manufacturer: FLIR Systems, Inc., Wilsonville, OR, USA) for infrared image acquisition at wavelength (7.5 µm < $\lambda$ < 13 µm). The resolution of the IR-camera was $640 \times 512$ pixels with field of view (FOV) of $45°(H) \times 37°(V)$ and a 13 mm focal length. This provides 1.31 m rad Spatial resolution (IFOV). The frame rate of the camera was able to reach 9Hz but it was 1 second interval time between the frames to avoid high computational complexity. The object temperature range can be varied from −40 °C to +160 °C. During the experiment the subjects were sitting at 1.2 m and lying at 3 m distance in front of the camera. The camera was located at 1.7 m from the ground with an angle of 36.5° from the vertical axis. Figure 3 shows the schematic experimental setup along with two examples of thermal images.

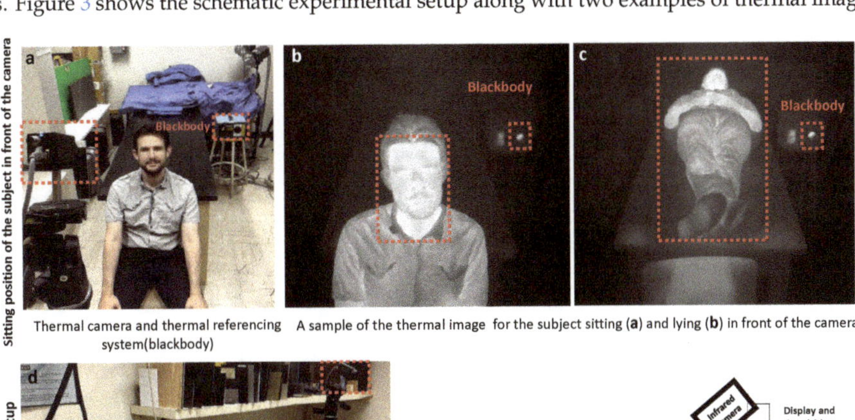


Thermal camera and thermal referencing system(blackbody)    A sample of the thermal image for the subject sitting (**a**) and lying (**b**) in front of the camera

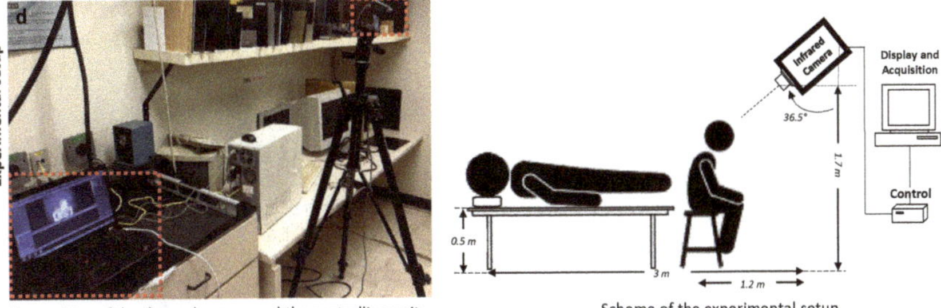


The position of the thermal camera and the controlling unit    Scheme of the experimental setup

**Figure 3.** The experimental setup of the approach is shown along with schematic design of the experiment. (**a**) shows an example of a subject sitting in the field of view of the IR-camera. (**b,c**) are the two thermal image samples of sitting and lying in front of the camera. (**d**) presents the infrared camera and the camera's interface on the computer during the experiment. The position of the blackbody and subject as our ROI are highlighted.

The thermal image database (The mentioned dataset is available for research uses. A sample set of this database is provided at the following link: http://vision.gel.ulaval.ca/~{}bardia/web%20page/ThermalDatabase.html) was created using 15 subjects for two minutes of acquisition while the subjects are sitting or lying in front of the camera. The subjects were free to have arbitrary movements during the acquisition to not only simulate the actual condition but also to verify the ability of the system to deal with motion artifacts. The processing was carried out with a PC (Intel(R) Core(TM) i7 CPU, 930, 2.80 GHz, RAM 24.00 GB, 64 bit Operating System) and processing of the thermal data was conducted using MATLAB computer program.

*3.2. System Evaluation*

To verify the performance of the system in the presence of noise, we added seven levels of additive Salt and pepper noise into the input stream. The system showed considerable robustness against noise due to the incremental low-rank noise reduction stage in the process. Figure 4 represents the performance of the system involving additive noise and computational complexity of the proposed approach. Figure 5 shows a participant during the simulation, during which is processed with Matlab while a heating source was attached to the volunteer's body. Figure 5a,b show two time points at start and middle of the experiment when the temperature of the heating source has not increased yet. Figure 5c represents segmented hot spots during imaging and their temperatures can be measured. Figure 6 also shows two examples of system performance for 2% and 20% noise. The ROIs for both cases are correctly found and tracked during the experiments.

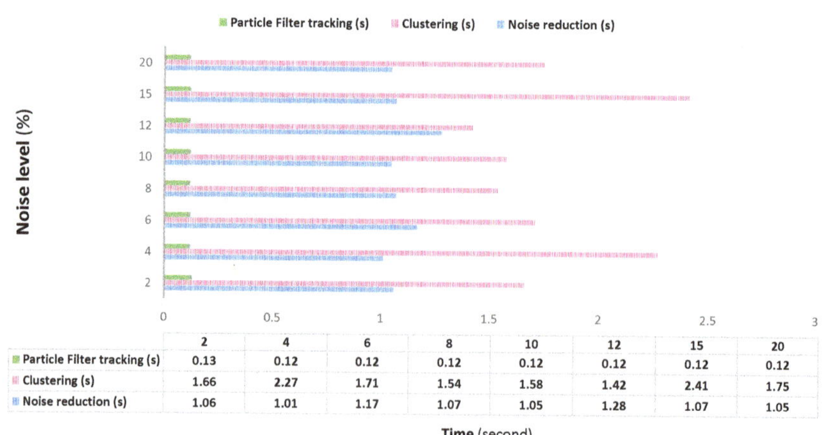


| | 2 | 4 | 6 | 8 | 10 | 12 | 15 | 20 |
|---|---|---|---|---|---|---|---|---|
| Particle Filter tracking (s) | 0.13 | 0.12 | 0.12 | 0.12 | 0.12 | 0.12 | 0.12 | 0.12 |
| Clustering (s) | 1.66 | 2.27 | 1.71 | 1.54 | 1.58 | 1.42 | 2.41 | 1.75 |
| Noise reduction (s) | 1.06 | 1.01 | 1.17 | 1.07 | 1.05 | 1.28 | 1.07 | 1.05 |

**Figure 4.** The robustness of the tracking approach is tested against the additive noise and computational load of this process is shown in the table and bar-plot from 2% to 20% additive noise.

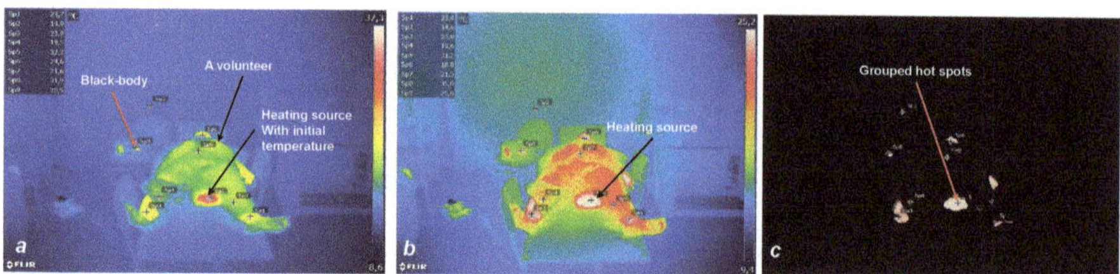


**Figure 5.** Thermal measurement and tracking conducted by an experiment. (**a**–**c**) are thermal images of a volunteer with its measuring temperature points before (**a**) and after (**b**) increasing the temperature of a synthetic elevating thermal source, and heating spot were tracked and clustered by our algorithm (**c**).

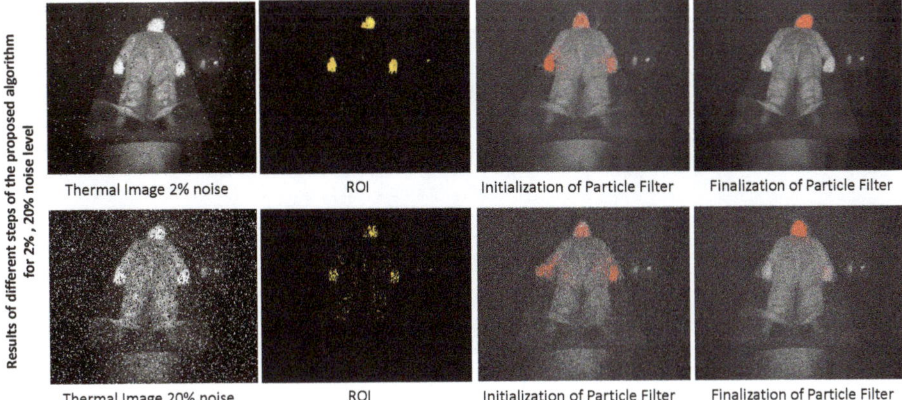



**Figure 6.** The results of the proposed approach are shown for two levels of noise (2% and 20%).

## 4. Discussion

This study has shown the application of infrared imagery in the 7.5–13 µm wavelength range for temperature monitoring using unsupervised learning techniques. One of the requirements of the proposed system is an initial calibration to obtain the real temperature. However this is not an issue due to the use of a black-body as a reference temperature for the healthy body temperature (around 36.5 °C). The system interestingly covers the particles of the regions where the highest temperature is located (which usually involves the face of the subjects). Following the results, the particle filter was initiated on the group which showed the higher temperature and segmented by clustering then shifted to the higher temperature points. That is because of the tendency of the particles in the particle filter to track the higher value points. This process usually involved some initial time and then stabilized to a certain region.

The other matter to be discussed is related to the noise reduction and the level of additive noise. The system has shown a considerable performance to track the thermal regions while the system was exposed to 2% to 20% additive salt&pepper noise and the results have provided a confirmation for this matter. The good performance of the system is a result of the low rank noise reduction which was carried out though an incremental Eigen-decomposition method (SVD). Incremental-SVD improved the heavy process of batch-SVD but it is still a heavy stage in the proposed algorithm and the system achieved a reasonable computational complexity due to down-sampling in the frame rate to compensate for the processing time.

The contributions of the proposed approach lie under two major points of view, i.e., a modification in applications of thermal object tracking and incremental noise reduction. In this application modification, we have modified the previous approach ([38]) by concentrating more on normal body temperature and adjusting this approach by using a black-body as a temperature reference. This provides a more applicable system for medical and health care usages. Moreover, the contribution involving low rank noise reduction has provided considerable novelty to the previous similar approaches such as [38,39,67].

## 5. Conclusions

The approach presented is an automatic monitoring system for the patient's body temperature using incremental low-rank noise reduction and applying Incremental SVD and applied color based K-means clustering to find the ROI and a particle filter to track the ROI(s) within the sequence. The system was tested in an experiment conducted to simulate the hospital's conditions which involved 15 subjects in sitting and lying positions. The robustness of the method was verified in the presence of subject's arbitrary movements and additive noise. The system was able to function even in the presence of 20% (salt & pepper) noise with reasonable computational load. As future work, the low ranking noise reduction process can be modified to avoid the vulnerability of the system when facing a higher level of noise. This objective can be reached by using an additional penalty term in the computing of the low rank data representation or by adding a possible relaxation coefficient into the calculation.

**Author Contributions:** Conceptualization, design and implementation B.Y. and H.M.S.; methodology, B.Y.; data-set curation, M.E., C.I.-C.; supervisor of M.E., D.L.; writing draft preparation, B.Y.; review and editing, all authors; main supervision, X.P.V.M.; project administration, C.I.-C., X.P.V.M., R.W. and M.K.; main PI and funding acquisition, X.P.V.M.

**Funding:** This research was conducted under Canadian-Tier 1 Research Chair in Multipolar Infrared Vision (MIVIM) at Computer Vision and Systems Laboratory (CVSL), Laval University.

**Acknowledgments:** The authors would like to acknowledge and thank Dr. Annette Schwerdtfeger from the Department of Electrical and Computer Engineering at Laval University for her constructive comments, proof reading, and help. Also, we would like thank anonymous reviewers for their insightful and constructive comments.

**Conflicts of Interest:** R.W. and M.K. are with RT thermal and Visiooimage companies, respectively. The rest of the authors declare no conflict of interest.

## Abbreviations

The following abbreviations are used in this manuscript:

| | |
|---|---|
| ROI | region of interest |
| SVD | singular value decomposition |
| FOV | field of view |
| 3D | three dimension |
| PCL | point cloud library |
| ICP | iterative closest points |

## References


1. Jones, B.F. A reappraisal of the use of infrared thermal image analysis in medicine. *IEEE Trans. Med. Imaging* **1998**, *17*, 1019–1027. [CrossRef] [PubMed]
2. Wu, D.; Lu, H.; Zhao, B.; Liu, J.; Zhao, M. Micro-Motion Dynamics and Shape Parameters Estimation Based on an Infrared Signature Model of Spatial Targets. *Electronics* **2019**, *8*, 755. [CrossRef]
3. Abdel-Nasser, M.; Moreno, A.; Puig, D. Breast Cancer Detection in Thermal Infrared Images Using Representation Learning and Texture Analysis Methods. *Electronics* **2019**, *8*, 100. [CrossRef]
4. Etehadtavakol, M.; Ng, E.Y. Potential of Thermography in Pain Diagnosing and Treatment Monitoring. In *Application of Infrared to Biomedical Sciences*; Springer: New York, NY, USA, 2017; pp. 19–32.
5. Manohar, S.; Vaartjes, S.E.; van Hespen, J.C.; Klaase, J.M.; van den Engh, F.M.; Steenbergen, W.; Van Leeuwen, T.G. Initial results of in vivo non-invasive cancer imaging in the human breast using near-infrared photoacoustics. *Opt. Express* **2007**, *15*, 12277–12285. [CrossRef] [PubMed]
6. Merla, D.A.; Tsiamyrtzis, P.; Pavlidis, I. Imaging Facial Signs of Neurophysiological Responses. *IEEE Trans. Biomed. Eng.* **2009**, *56*, 477–484.
7. Tabatabaei, N.; Mandelis, A.; Amaechi, B.T. Thermophotonic lock-in imaging of early demineralized and carious lesions in human teeth. *J. Biomed. Opt.* **2011**, *16*, 071402. [CrossRef]
8. Chiang, M.F.; Lin, P.W.; Lin, L.F.; Chiou, H.Y.; Chien, C.W.; Chu, S.F.; Chiu, W.T. Mass screening of suspected febrile patients with remote-sensing infrared thermography: Alarm temperature and optimal distance. *J. Formos. Med. Assoc.* **2008**, *107*, 937–944. [CrossRef]
9. Dempsey, M.F.; Condon, B.; Hadley, D.M. Investigation of the factors responsible for burns during MRI. *J. Magn. Reson. Imaging Off. J. Int. Soc. Magn. Reson. Med.* **2001**, *13*, 627–631. [CrossRef]
10. Shellock, F.G.; Kanal, E. Burns associated with the use of monitoring equipment during MR procedures. *J. Magn. Reson. Imaging* **1996**, *6*, 271–272. [CrossRef]

11. Brown, T.; Goldstein, B.; Little, J. Severe burns resulting from magnetic resonance imaging with cardiopulmonary monitoring. Risks and relevant safety precautions. *Am. J. Phys. Med. Rehabi.* **1993**, *72*, 166–167.

12. Bashein, G.; Syrovy, G. Burns associated with pulse oximetry during magnetic resonance imaging. *Anesthesiol. J. Am. Soc. Anesthesiol.* **1991**, *75*, 382. [CrossRef] [PubMed]

13. Hall, S.C.; Stevenson, G.; Suresh, S. Burn associated with temperature monitoring during magnetic resonance imaging. *Anesthesiol. J. Am. Soc. Anesthesiol.* **1992**, *76*, 152. [CrossRef] [PubMed]

14. Albuquerque, R.J.; de Leeuw, R.; Carlson, C.R.; Okeson, J.P.; Miller, C.S.; Andersen, A.H. Cerebral activation during thermal stimulation of patients who have burning mouth disorder: An fMRI study. *Pain* **2006**, *122*, 223–234. [CrossRef] [PubMed]

15. Lemieux, L.; Allen, P.J.; Franconi, F.; Symms, M.R.; Fish, D.K. Recording of EEG during fMRI experiments: Patient safety. *Magn. Reson. Med.* **1997**, *38*, 943–952. [CrossRef]

16. Simi, S.; Ballardin, M.; Casella, M.; De Marchi, D.; Hartwig, V.; Giovannetti, G.; Vanello, N.; Gabbriellini, S.; Landini, L.; Lombardi, M. Is the genotoxic effect of magnetic resonance negligible? Low persistence of micronucleus frequency in lymphocytes of individuals after cardiac scan. *Mutat. Res./Fund. Mol. Mech. Mutagenesis* **2008**, *645*, 39–43. [CrossRef]

17. Bonassi, S.; Znaor, A.; Ceppi, M.; Lando, C.; Chang, W.P.; Holland, N.; Kirsch-Volders, M.; Zeiger, E.; Ban, S.; Barale, R.; et al. An increased micronucleus frequency in peripheral blood lymphocytes predicts the risk of cancer in humans. *Carcinogenesis* **2007**, *28*, 625–631. [CrossRef]

18. Hartwig, V.; Giovannetti, G.; Vanello, N.; Lombardi, M.; Landini, L.; Simi, S. Biological effects and safety in magnetic resonance imaging: A review. *Int. J. Environ. Res. Public Health* **2009**, *6*, 1778–1798. [CrossRef]

19. Jin, J. *Electromagnetic Analysis and Design in Magnetic Resonance Imaging*; Routledge: London, UK, 2018.

20. Pykett, I.L. NMR imaging in medicine. *Sci. Am.* **1982**, *246*, 78–91. [CrossRef]

21. Rodgers, P.M.; Ward, J.; Baudouin, C.J.; Ridgway, J.P.; Robinson, P.J. Dynamic contrast-enhanced MR imaging of the portal venous system: Comparison with x-ray angiography. *Radiology* **1994**, *191*, 741–745. [CrossRef]

22. Laakman, R.; Kaufman, B.; Han, J.; Nelson, A.; Clampitt, M.; O'Block, A.; Haaga, J.; Alfidi, R. MR imaging in patients with metallic implants. *Radiology* **1985**, *157*, 711–714. [CrossRef]

23. Morgan, D.E.; Kenney, P.J.; Meeks, M.C.; Pile, N.S. MR imaging of breast implants and their complications. *AJR Am. J. Roentgenol.* **1996**, *167*, 1271–1275. [CrossRef] [PubMed]

24. Formica, D.; Silvestri, S. Biological effects of exposure to magnetic resonance imaging: An overview. *Biomed. Eng. Online* **2004**, *3*, 11. [CrossRef] [PubMed]

25. Shellock, F.G. Radiofrequency energy-induced heating during MR procedures: A review. *J. Magn. Reson. Imaging* **2000**, *12*, 30–36. [CrossRef]

26. Shellock, F.G.; Rothman, B.; Sarti, D. Heating of the scrotum by high-field-strength MR imaging. *AJR Am. J. Roentgenol.* **1990**, *154*, 1229–1232. [CrossRef]

27. Shellock, F.; Crues, J.V. Corneal temperature changes induced by high-field-strength MR imaging with a head coil. *Radiology* **1988**, *167*, 809–811. [CrossRef] [PubMed]

28. Vahlensieck, M. Tattoo-related cutaneous inflammation (burn grade I) in a mid-field MR scanner. *Eur. Radiol.* **2000**, *10*, 197. [CrossRef] [PubMed]

29. Wagle, W.A.; Smith, M. Tattoo-induced skin burn during MR imaging. *Am. J. Roentgenol.* **2000**, *174*, 1795. [CrossRef]

30. Kreidstein, M.; Giguere, D.; Freiberg, A. MRI interaction with tattoo pigments: Case report, pathophysiology, and management. *Plast. Reconstr. Surg.* **1997**, *99*, 1717–1720. [CrossRef]

31. Cleveland, R.F., Jr.; Athey, T.W. Specific absorption rate (SAR) in models of the human head exposed to hand-held UHF portable radios. *Bioelectromagn. J. Bioelectromagn. Soc. Soc. Phys. Regul. Biol. Med. Eur. Bioelectromagn. Assoc.* **1989**, *10*, 173–186. [CrossRef]

32. International Electrotechnical Commission. *Particular Requirements for the Basic Safety and Essential Performance of Magnetic Resonance Equipment for Medical Diagnosis*; Technical Report; International Electrotechnical Commission (IEC): Geneva, Switzerland, 2010; p. 60601-2-33.

33. Adair, E.R.; Berglund, L.G. On the thermoregulatory consequences of NMR imaging. *Magn. Reson. Imaging* **1986**, *4*, 321–333. [CrossRef]

34. Shellock, F.G.; Schaefer, D.J.; Crues, J.V. Alterations in body and skin temperatures caused by MRI. *Br. J. Radiol.* **1990**, *63*, 317. [CrossRef] [PubMed]

35. Van den Berg, C.A.; Van den Bergen, B.; Van de Kamer, J.B.; Raaymakers, B.W.; Kroeze, H.; Bartels, L.W.; Lagendijk, J.J. Simultaneous B homogenization and specific absorption rate hotspot suppression using a magnetic resonance phased array transmit coil. *Magn. Reson. Med. Off. J. Int. Soc. Magn. Reson. Med.* **2007**, *57*, 577–586. [CrossRef] [PubMed]

36. Liu, F.; Zhao, H.; Crozier, S. Calculation of electric fields induced by body and head motion in high-field MRI. *J. Magn. Reson.* **2003**, *161*, 99–107. [CrossRef]

37. Stathopoulos, I.; Skouroliakou, K.; Michail, C.; Valais, I. Dynamic infrared thermography study of blood flow relative to lower limp position. In *Journal of Physics: Conference Series*; IOP Publishing: Bristol, UK, 2015; Volume 637, p. 012027.

38. Yousefi, B.; Fleuret, J.; Zhang, H.; Maldague, X.P.; Watt, R.; Klein, M. Automated assessment and tracking of human body thermal variations using unsupervised clustering. *Appl. Opt.* **2016**, *55*, D162–D172. [CrossRef]

39. Bilodeau, G.A.; Torabi, A.; Lévesque, M.; Ouellet, C.; Langlois, J.P.; Lema, P.; Carmant, L. Body temperature estimation of a moving subject from thermographic images. *Mach. Vis. Appl.* **2012**, *23*, 299–311. [CrossRef]

40. Kennedy, H.V. Modeling noise in thermal imaging systems. *Proc. SPIE* **1993**, *1969*, 66–70.

41. Watts, M.R.; Shaw, M.J.; Nielson, G.N. Optical resonators: Microphotonic thermal imaging. *Nat. Photonics* **2007**, *1*, 632–634. [CrossRef]

42. Cardone, D.; Pinti, P.; Merla, A. Thermal infrared imaging-based computational psychophysiology for psychometrics. *Comput. Math. Methods Med.* **2015**, *2015*. [CrossRef]

43. Manabe, M.; Yamazaki, H.; Sakai, K. Shape factor simulation for the thermal radiation environment of the human body and the VRML visualization. *Build. Environ.* **2004**, *39*, 927–937. [CrossRef]

44. Badel, E.; Delisee, C.; Lux, J. 3D structural characterisation, deformation measurements and assessment of low-density wood fibreboard under compression: The use of X-ray microtomography. *Compos. Sci. Technol.* **2008**, *68*, 1654–1663. [CrossRef]

45. Jaganathan, S.; Tafreshi, H.V.; Pourdeyhimi, B. Modeling liquid porosimetry in modeled and imaged 3-D fibrous microstructures. *J. Colloid Interface Sci.* **2008**, *326*, 166–175. [CrossRef] [PubMed]

46. Cheng, V.S.; Bai, J.; Chen, Y. A high-resolution three-dimensional far-infrared thermal and true-color imaging system for medical applications. *Med. Eng. Phys.* **2009**, *31*, 1173–1181. [CrossRef] [PubMed]

47. Sharma, R.; Locke, B.R. Jet fuel toxicity: Skin damage measured by 900-MHz MRI skin microscopy and visualization by 3D MR image processing. *Magn. Reson. Imaging* **2010**, *28*, 1030–1048. [CrossRef] [PubMed]

48. Markelj, P.; Tomaževič, D.; Likar, B.; Pernuš, F. A review of 3D/2D registration methods for image-guided interventions. *Med. Image Anal.* **2012**, *16*, 642–661. [CrossRef]

49. Lagüela, S.; Armesto, J.; Arias, P.; Herráez, J. Automation of thermographic 3D modelling through image fusion and image matching techniques. *Autom. Constr.* **2012**, *27*, 24–31. [CrossRef]

50. Schutt, D.J.; Swindle, M.M.; Helke, K.L.; Bastarrika, G.; Schwarz, F.; Haemmerich, D. Sequential activation of ground pads reduces skin heating during radiofrequency tumor ablation: In vivo porcine results. *IEEE Trans. Biomed. Eng.* **2009**, *57*, 746–753. [CrossRef]

51. Vidas, S.; Moghadam, P. HeatWave: A handheld 3D thermography system for energy auditing. *Energy Build.* **2013**, *66*, 445–460. [CrossRef]

52. Siewert, C.; Dänicke, S.; Kersten, S.; Brosig, B.; Rohweder, D.; Beyerbach, M.; Seifert, H. Difference method for analysing infrared images in pigs with elevated body temperatures. *Zeitschrift für Medizinische Physik* **2014**, *24*, 6–15. [CrossRef]

53. Bay, H.; Tuytelaars, T.; Van Gool, L. Surf: Speeded up robust features. In *European Conference on Computer Vision*; Springer: New York, NY, USA, 2006; pp. 404–417.

54. Chen, Y.; Medioni, G. Object modelling by registration of multiple range images. *Image Vis. Comput.* **1992**, *10*, 145–155. [CrossRef]

55. Rusu, R.B.; Cousins, S. Point cloud library (pcl). In *Proceedings of the 2011 IEEE International Conference on Robotics and Automation*, Shanghai, China, 2011; pp. 1–4.

56. Golub, G.H.; Van Loan, C.F. *Matrix computations*; JHU Press: Baltimore, MD, USA, 2012; Volume 3.

57. Bunch, J.R.; Nielsen, C.P. Updating the singular value decomposition. *Numer. Math.* **1978**, *31*, 111–129. [CrossRef]

58. Brand, M. Incremental singular value decomposition of uncertain data with missing values. In *Computer Vision—ECCV 2002*; Springer: Berlin/Heidelberg, Germany, 2002; pp. 707–720.

59. Jain, A.K. Data clustering: 50 years beyond K-means. *Pattern Recognit. Lett.* **2010**, *31*, 651–666. [CrossRef]

60. Yousefi, P.; Jalab, H.; Ibrahim, R.; Mohd Noor, N.; Ayub, M.; Gani, A. River segmentation using satellite image contextual information and Bayesian classifier. *Imaging Sci. J.* **2016**, *64*, 453–459. [CrossRef]

61. Isard, M.; Blake, A. Condensation—Conditional density propagation for visual tracking. *Int. J. Comput. Vis.* **1998**, *29*, 5–28. [CrossRef]

62. Hoseini, S.; Kabiri, P. A Novel Feature-Based Approach for Indoor Monocular SLAM. *Electronics* **2018**, *7*, 305. [CrossRef]

63. Zhang, F.; Hancock, E.R.; Goodlett, C.; Gerig, G. Probabilistic white matter fiber tracking using particle filtering and von Mises–Fisher sampling. *Med. Image Anal.* **2009**, *13*, 5–18. [CrossRef] [PubMed]

64. Smal, I.; Meijering, E.; Draegestein, K.; Galjart, N.; Grigoriev, I.; Akhmanova, A.; Van Royen, M.; Houtsmuller, A.B.; Niessen, W. Multiple object tracking in molecular bioimaging by Rao-Blackwellized marginal particle filtering. *Med. Image Anal.* **2008**, *12*, 764–777. [CrossRef] [PubMed]

65. Smal, I.; Meijering, E. Quantitative comparison of multiframe data association techniques for particle tracking in time-lapse fluorescence microscopy. *Med. Image Anal.* **2015**, *24*, 163–189. [CrossRef]

66. Pitt, M.K.; Shephard, N. Filtering via simulation: Auxiliary particle filters. *J. Am. Stat. Assoc.* **1999**, *94*, 590–599. [CrossRef]

67. Bilodeau, G.A.; Ghali, R.; Desgent, S.; Langlois, P.; Farah, R.; St-Onge, P.L.; Duss, S.; Carmant, L. Where is the rat? Tracking in low contrast thermographic images. In Proceedings of the 2011 IEEE Computer Society Conference on Computer Vision and Pattern Recognition Workshops (CVPRW), Colorado Springs, CO, USA, 2011; pp. 55–60.

 *electronics*

*Article*

# Soft Elbow Exoskeleton for Upper Limb Assistance Incorporating Dual Motor-Tendon Actuator


**Rifky Ismail [1], Mochammad Ariyanto [1], Inri A. Perkasa [1], Rizal Adririanto [1], Farika T. Putri [1,2], Adam Glowacz [3] and Wahyu Caesarendra [4,*]**

[1]  Center for Biomechanics, Biomaterial, Biomechatronics, and Biosignal Processing (CBIOM3S), Diponegoro University, Semarang 50275, Indonesia; rifky_ismail@ft.undip.ac.id (R.I.); mochammad_ariyanto@ft.undip.ac.id (M.A.); adipandiangan13@gmail.com (I.A.P.); rizaladrianto@gmail.com (R.A.); farikatonoputri@gmail.com (F.T.P.)
[2]  Semarang State Polytechnic, Jalan Prof. Soedarto, SH. Tembalang, Semarang, Central Java 50275, Indonesia
[3]  Department of Automatic Control and Robotics, AGH University of Science and Technology, Al. A. Mickiewicza 30, 30-059 Kraków, Poland; adglow@agh.edu.pl
[4]  Faculty of Integrated Technologies, Universiti Brunei Darussalam, Jalan Tungku Link, Gadong BE1410, Brunei
*  Correspondence: wahyu.caesarendra@ubd.edu.bn; Tel.: +673-7345-623



Received: 6 September 2019; Accepted: 14 October 2019; Published: 18 October 2019



**Abstract:** Loss of muscle functions, such as the elbow, can affect the quality of life of a person. This research is aimed at developing an affordable two DOF soft elbow exoskeleton incorporating a dual motor-tendon actuator. The soft elbow exoskeleton can be used to assist two DOF motions of the upper limb, especially elbow and wrist movements. The exoskeleton is developed using fabric for the convenience purpose of the user. The dual motor-tendon actuator subsystem employs two DC motors coupled with lead-to-screw converting motion from angular into linear motion. The output is connected to the upper arm hook on the soft exoskeleton elbow. With this mechanism, the proposed actuator system is able to assist two DOF movements for flexion/extension and pronation/supination motion. Proportional-Integral (PI) control is implemented for controlling the motion. The optimized value of Kp and Ki are 200 and 20, respectively. Based on the test results, there is a slight steady-state error between the first and the second DC motor. When the exoskeleton is worn by a user, it gives more steady-state errors because of the load from the arm weight. The test results demonstrate that the proposed soft exoskeleton elbow can be worn easily and comfortably by a user to assist two DOF for elbow and wrist motion. The resulted range of motion (ROM) for elbow flexion–extension can be varied from 90° to 157°, whereas the maximum of ROM that can be achieved for pronation and supination movements are 19° and 18°, respectively.

**Keywords:** elbow soft exoskeleton; dual motor-tendon actuator; PI control; flexion/extension; pronation/supination


---

## 1. Introduction

Wearable robot exoskeleton can be used as a device to facilitate the user in carrying out an activity or work. The exoskeleton robot can be used in the medical field as a healing therapeutic device for people with paralysis due to stroke or paralysis due to accident, which is often called brachialis plexus injury (BPI). Wearable exoskeletons are a very good development for people with disabilities and provide mechanical assistance in their activities.

In terms of the material used, the exoskeleton robot is divided into two classes: Hard exoskeleton and soft exoskeleton. A hard exoskeleton is widely used when high mechanical force, precise position, and fast dynamics are required [1]. On the other hand, a soft exoskeleton is utilized when it needs portability and comfortability for people who wear the exoskeleton robot. In order to enhance the performance of the exoskeleton, some researchers combine these two kinds of exoskeleton robots.

Researchers have developed a hard exoskeleton elbow for assisting people with the upper limb [2–8]. They succeeded in building a hard elbow exoskeleton robot to provide mechanical assistance for a user/wearer mostly for flexion and extension motion. A DC motor is widely used for both elbow and hand exoskeletons. By applying this hard robot mechanism, high force and torque can be achieved to provide mechanical support for the wearer. Surface electromyography is a widely used sensor in exoskeletons. The surface electromyography sensor functions as interface between the healthy muscle and the exoskeleton [9–12].

Soft robotics is one of the areas of robotic research aimed to solve a challenge faced by traditional robotics. Wearable exoskeleton in this research is part of soft robotic area, which consists of manufactured flexible structure [13]. There are some benefits gained from the use of a wearable exoskeleton robot as health assisting device: A soft exoskeleton tends to be lightweight, thus, people can comfortably use it, and, most importantly, the soft exoskeleton is flexible and can therefore accomplish difficult motions that a traditional robotic can not [13,14].

Actuator design plays an important role in soft exoskeleton development. A soft robotic elbow sleeve designed by a researcher from the National University of Singapore used the combination of flexion and extension actuator. Flexion actuator is being forced into a stiff fixture with the intention of nonlinear effect minimization. An experiment is conducted to test the deflection of the extension actuator. Weight variation is attached at one side of the extension actuator. Meanwhile, the other end is being put to fix [13]. Soft robotic wearable elbow exoskeleton made by researchers from Carlos III Univesity of Madrid, Spain, used a shape memory actuator (SMA) with some advantages, e.g., lightweight, minimal noise when operated, and low cost to be produced. SMA actuator transducer material uses wires or springs to convert thermal energy into mechanical energy. Electric current circulates through SMA, causing a heated effect in SMA. Electric energy is converted into mechanical energy in SMA, called the Joule effect. The thermal energy leads to the SMA element recovery process to its original form. The SMA's recovery energy then being transformed into mechanical work [15].

One of the most widely used actuators in the soft exoskeleton is Bowden cable connected to a DC motor. A researcher from the Czech Technical University developed a smart upper limb orthosis utilizing Bowden cable attached to one end of a limb and a motor stepper to another end [16]. Another upper limb exoskeleton employing the Bowden cable transmission in their actuator is made by some researchers from China [17]. In this study, Bowden cable is used to minimize the man–machine interaction force. Bowden cable is commonly used as a transmission force in an elbow and hand exoskeleton [18–22]. Some researchers have conducted a study in combining pneumatic with Bowden cable [23,24] for the exoskeleton actuator system.

In this study, an affordable and lightweight soft elbow exoskeleton incorporating dual motor-tendon actuation is developed. The actuation system consists of a Bowden cable connected to a lead screw, which is coupled with a DC motor in order to move the elbow or wrist. An infrared sensor is placed in the actuator system to measure the obstacle movement. This non-back drivable mechanism enables the proposed soft elbow exoskeleton to provide mechanical support/force for the elbow to move the user's elbow and wrist in two DOF, i.e., flexion/extension and pronation/supination. This actuation system is easy to manufacture and the components are widely available in online and offline stores. Proportional-Integral (PI) control is utilized for controlling the linear displacement of the motor-tendon actuation for pulling/pushing the elbow and wrist. PID and PI compensators are widely used in controlling the motion of both hard and soft exoskeletons [7,13,25,26]. The proposed soft elbow exoskeleton is tested by a user to provide the mechanical force for flexion/extension and pronation/supination movements.

Most of the elbow exoskeletons resulted from previous research. Both hard and soft elbow exoskeletons can provide mechanical support for flexion and extension movements without assisting the pronation and supination motion [7,12,13,22,24–26]. Therefore, we propose an affordable dual motor-tendon actuator that can provide mechanical force for flexion/extension and pronation/supination movements. The flexion/extension motion can be attained by pulling/pushing the two Bowden cables simultaneously, whereas the pronation/supination movement can be obtained by pulling and pushing the two Bowden cables in the opposite direction.

There are some advantages of the aforementioned exoskeleton compared to other soft and hard exoskeleton designs that have been studied. User comfortability is one of the concerns in producing this soft exoskeleton which is achieved through the usage of fabric that is delicate yet has the ability to assist in supporting mechanical force and support. It has a simple method for fixing, alignment, and is easy to use for a user. The soft exoskeleton uses affordable materials and spare parts are easy to find, which can be beneficial to the exoskeleton user when there is any need for replacement during exoskeleton usage. As a result, the soft elbow exoskeleton can be affordable. The non-backdriveable mechanism allows holding/maintaining a user's hand during the flexion/extension movement without consuming more electric power from the battery. This indicates that the proposed exoskeleton is an energy-efficient device. The motor-tendon actuation system enables to convert motion directly from linear displacement from the actuator output to flexion/extension and pronation/supination motion on the user's arm. This also increases the safety of a user when the exoskeleton is worn and operated. The challenge for this exoskeleton is to provide a range of motion ability for flexion/extension and pronation/supination. The mechanism requires more space in the actuator system for a longer range of motion (ROM).

## 2. Soft Elbow Exoskeleton and Motor-Tendon Actuator Design

A user has to wear the elbow exoskeleton comfortably. Based on this consideration, the fabric is selected as a material of the proposed soft exoskeleton elbow which is worn by a user. Bowden cable is chosen as a force transmission in the soft exoskeleton actuation system. The proposed exoskeleton elbow is equipped with an upper vest exoskeleton suit. Figure 1 below is a soft elbow exoskeleton with two DOF in elbow and wrist movement. Two motor-tendon actuators are implemented for controlling each DOF. The first DOF is for flexion and extension motion while the second DOF is for pronation and supination movements.

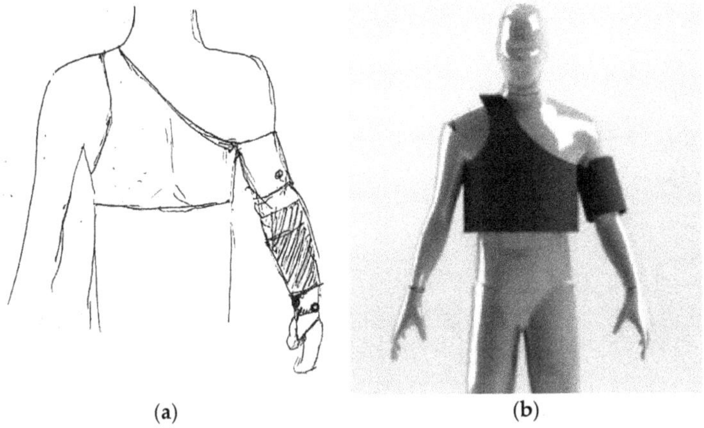

(a)       (b)

**Figure 1.** Soft elbow exoskeleton (**a**) proposed 2D design; (**b**) proposed 3D design in computer-aided design (CAD) software.

### 2.1. Fabric-Based Soft Elbow Exoskeleton Design

At this stage, 3D modeling of the robot elbow exoskeleton is conducted according to the dimensions of the arm of the patients with brachialis plexus injury. Modeling is conducted by developing a 2D design and modeling them into a 3D model using SolidWorks computer-aided design (CAD) software. The modeled component is a vest, wrist strap, and upper arm hook designed based on the average size of the Indonesian people. Figure 1 shows the resulted design of the soft elbow exoskeleton which is worn by a user. The hook design in the upper arm and wrist is presented in Figure 2.

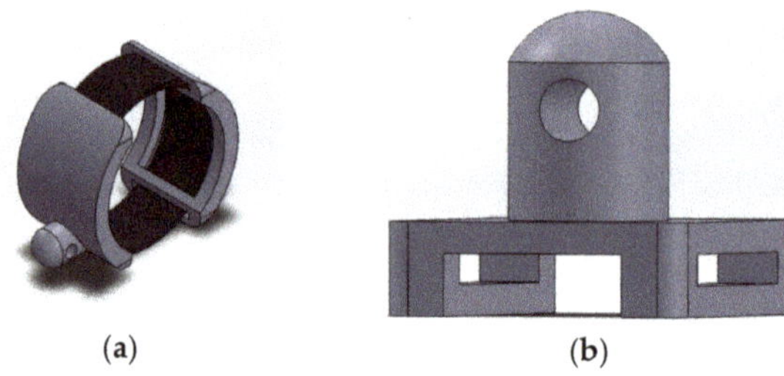

(a)        (b)

**Figure 2.** Hook design for soft exoskeleton elbow. (**a**) Wrist hook; (**b**) upper arm hook.

*2.2. Motor-Tendon Actuation Design*

In this study, actuator motor-tendon was chosen because of the ease in the manufacturing process which only requires a motor and a tendon to pull and stretch/push the hook on the upper arm and wrist. The motor used is a DC motor, and the Bowden cable is utilized as a tendon in this actuator. This wire/tendon is used as a connector between the arm hook and the nut that is on the lead screw on the actuator to move the elbow. In addition, an important component used is the infrared sensor which functions as a proximity sensor to provide feedback on the control system that is applied. The sensor is employed to measure an obstacle that represents the displacement or length of pull/push performed by the DC motor. The illustration of the motor-tendon actuator mechanism is shown in Figure 3. Limit switches are applied to limit the number of rotation of the DC motor.

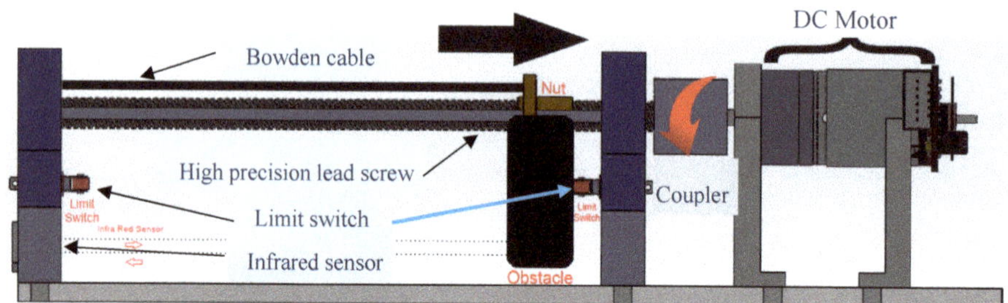


**Figure 3.** The mechanism of the proposed motor-tendon actuator.

2.2.1. Components

High precision lead screw with 8 mm in diameter and 150 mm in length is assembled to a nut serving as Bowden cable puller that is connected to the arm hook of the soft exoskeleton and driving the flexion and extension motion. A coupler is placed between the lead screw and DC motor to stretch the Bowden cable which acts as tendon which is connected to the arm. In this study, a DC motor with a JGA25-370 encoder is chosen and a L298N type motor driver is employed to regulate the direction and speed of the rotation of the motor. Arduino MEGA is used as the center of embedded control and data acquisition (DAQ) through serial communication.

SHARP GP2Y0A51SK0F infrared sensor is selected to measure the linear displacement of the Bowden cable, which will pull/push the user's arm. The sensor is able to measure at a distance from 2 cm to 15 cm with a size of 29.5 × 13.0 × 21.5 mm with analog output voltage. This range is suitable for the proposed motor-tendon actuation system. All mechanical and electrical components are placed strategically in actuator case and housing in order to produce an efficient and neat case design as well as comfortable to use. The actuator is powered by three 18650 Li-ion batteries with 3000 mAh and 3.7 V.

### 2.2.2. Infrared Sensor Calibration and Filter

The output voltage from the infrared sensor needs to be converted into a suitable signal (displacement of the actuator). Several signal acquisitions are recorded in the voltage read in terms of analog to digital converter (ADC). The polynomial equation obtained from third order polynomial curve fitting is expressed in (1).

$$y = -3 \times 10^{-7}x^3 + 0.0003x^2 - 0.1193x + 19.006 \tag{1}$$

The result from the obtained polynomial equation is applied to software processing. Because the infrared sensor produces noise signals in high frequency, a low-pass filter is employed in the filtering stage with the selected transfer function as expressed in (2). The measured infrared output voltage and the obtained third order polynomial are shown in Figure 4.

$$H(s) = \frac{1}{0.1\,s + 1} \tag{2}$$

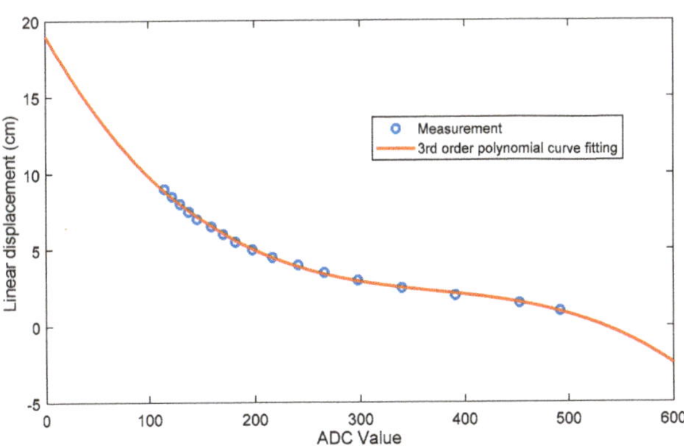


**Figure 4.** Obtained distance vs. analog to digital converter (ADC) infrared sensor output in the actuator system.

## 3. Two DOF Soft Elbow Exoskeleton and Control

### 3.1. Wearable Soft Elbow Exoskeleton System

When the soft exoskeleton is worn, the position of the attachment of the arm connected to the vest on the user's chest must be considered because it will affect the desired movement. In addition, the actuator case position must also be treated carefully because it is very influential for comfort in the use of soft exoskeleton. Figure 5 is an image of the 3D model of soft exoskeleton elbow, which was assembled and worn by a user.

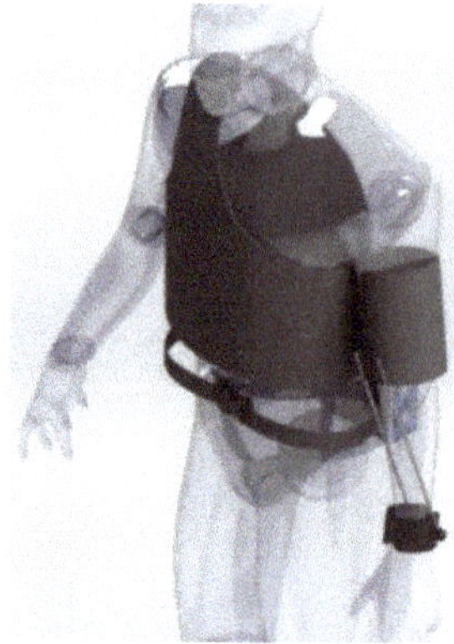

**Figure 5.** Final assembly of the proposed soft elbow exoskeleton.

### 3.1.1. Vest, Arm, and Wrist Hook

Figure 6a,b is the result of a prototype of a vest and an overall arm hook when it is worn by a user. Two Bowden cables are connected from the wrist hook to the two DC motors trough vest. The DC motors can rotate independently and allows the wrist hook to provide mechanical force for flexion/extension and pronation/supination. The total mass of the resulted soft elbow exoskeleton including arm and wrist hook is 358 g.

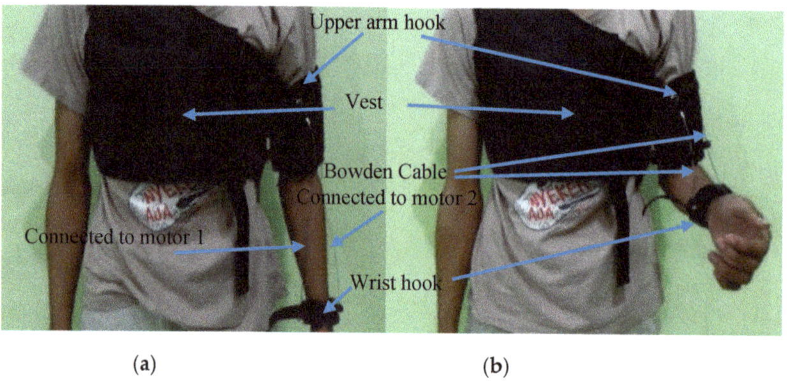


**Figure 6.** Final prototype of the proposed soft exoskeleton elbow (a) flexion; (b) extension

### 3.1.2. Motor-Tendon Actuator with Case

In this study, actuator motor-tendon incorporating lead screw is preferred to the soft exoskeleton because it can produce force high enough to assist flexion/extension and pronation/supination motion. Figure 7 reveals the result of the proposed motor-actuator tendon case. This actuator has several components where the function is to move the elbow soft exoskeleton on the user's arm. Several components contained in the actuator case can be seen clearly in Figure 7, and the parts of its components are described as in Table 1. The resulted prototype of the motor-tendon actuation system with the case design is presented in Figure 8. The dual motor-tendon actuator has a mass of 1655 g.

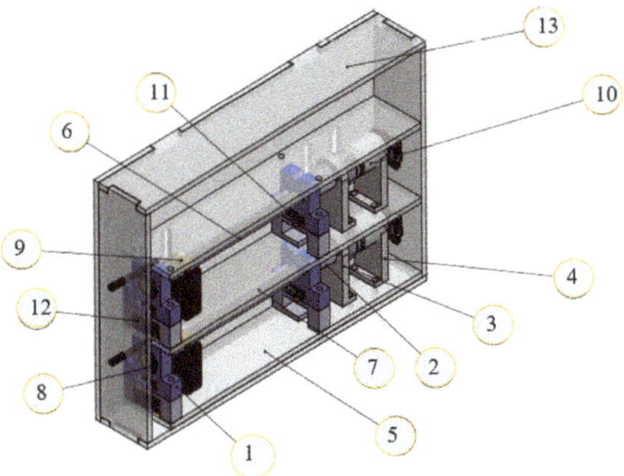

**Figure 7.** The dual motor-tendon actuator system.

**Table 1.** Proposed actuator components.

| No | Dual Motor-Tendon Component | Total |
|---|---|---|
| 1 | Bearing stand | 4 |
| 2 | Coupler | 2 |
| 3 | Upper/first motor stand | 2 |
| 4 | Lower/second motor stand | 2 |
| 5 | Base | 2 |
| 6 | Push Rod | 4 |
| 7 | Lead screw | 2 |
| 8 | Bearing | 2 |
| 9 | Nut | 2 |
| 10 | Motor DC | 2 |
| 11 | Limit Switch | 4 |
| 12 | Infrared Sensor | 2 |
| 13 | Case | 1 |

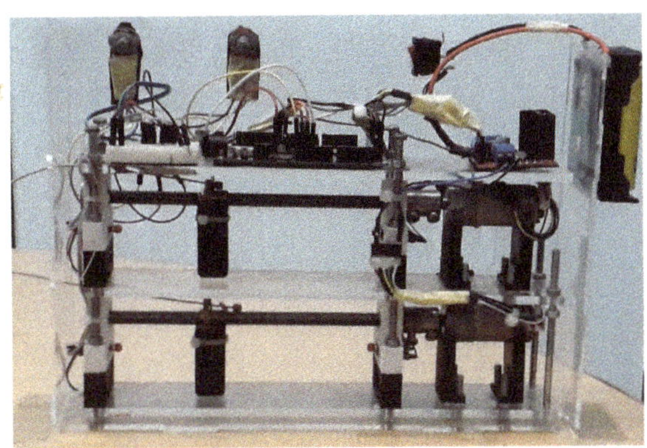

**Figure 8.** The result of dual motor-tendon actuator system.

## 3.2. PI Control

### 3.2.1. PID Tuner

Auto-tuning process is performed to identify the transfer function from the plant which is implemented in the PID tuner. It is utilized to identify the transfer function parameters for both motor-tendons. The parameters obtained are in the forms of K and $\tau$ (time constant) values. The transfer function parameters of both motors are identified according to the best result from the experiments that have been carried out. In this study, a PWM signal in the form of a step signal with a final value of 100 is employed as the input signal. The output signal is the linear displacement of Bowden cable measured by an infrared sensor. The acquired input and output signals are processed using the Matlab plant identification toolbox. The measurement and identification are performed twice to the exoskeleton when it is attached or not attached to the user.

From the identification of the plant that has been performed by using the MATLAB software, the obtained transfer function for the first and second motors in the form of second-order plant are expressed in Equations (3) and (4).

$$G_m(s) = \frac{1.2}{0.63\,s^2 + 17\,s} \tag{3}$$

$$G_m(s) = \frac{3.2}{1.9044\,s^2 + 45\,s} \tag{4}$$

Plant identification is also applied when the exoskeleton is paired/worn with the user. The user in this study is a person who has a normal healthy arm and hand. From the identification of the plant that has been conducted, the attained transfer function of the first and the second motor can be approached by a second-order plant as written in Equations (5) and (6).

$$G_m = \frac{0.1464}{0.36\,s^2 + 15\,s} \tag{5}$$

$$G_m = \frac{1.5}{1.9881\,s^2 + 35\,s} \tag{6}$$

### 3.2.2. Proportional-Integral (PI) Control

In this study, proportional-integral (PI) control aims to obtain better and appropriate position control for assisting the elbow flexion/extension and pronation/supination movements. The command used in this study is a potentiometer and programmed automatic signal as the input position/angle command and a proximity sensor to determine the displacement of the wire to move the arm while the motor as an output pulls or extends the Bowden cables. To obtain the optimal PI value, tests are performed when the exoskeleton is worn and not worn by a user. These tests are conducted to acquire the proportional constant (Kp), and the Integral constant (Ki) which are the most appropriate so that the control runs well between inputs and outputs for both with or without a user. The basic equation of PI control is expressed in Equation (7). The Simulink PI control block which is used in this study can be seen in Figure 9, while the embedded control block of controlling the soft elbow exoskeleton with PI control is presented in Figure 10. This block diagram is embedded into an Arduino Mega microcontroller.

$$PWM = Kp\,e(t) + \int_0^t Ki\,e(t)dt \tag{7}$$

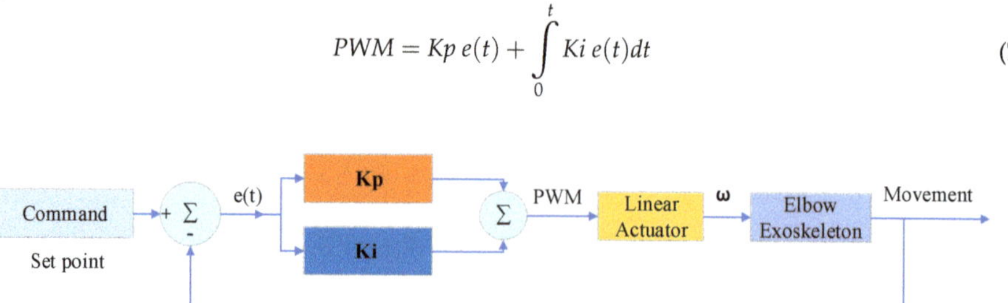


**Figure 9.** Proportional-Integral (PI) control block for the motor-tendon actuator.

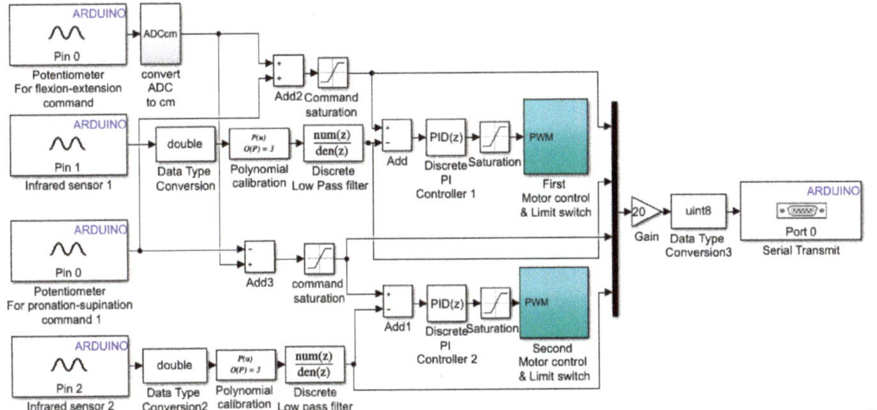


**Figure 10.** Embedded PI controller for controlling the dual motor-tendon actuator.

To obtain the optimum PI values, experiments are performed several times so that the best values of Kp and Ki are acquired. The tuning process of the PI parameters is easier to conduct in Simulink embedded control using Simulink Support Package for Arduino Hardware. The obtained parameters of the PI compensator should have a good response and precision control, which the output signal is able to follow the input signal well, and has a low steady-state error, less time constant, and a fast response.

## 4. Result and Discussion

The constant parameters of Kp and Ki are acquired based on the experimental works when the exoskeleton is attached and not attached to the user. A study participant with a normal healthy hand and arm at the age of 24 was selected for this test. The exoskeleton was designed with a range of motion (ROM) for flexion/extension of up to 70 degrees. This ROM was chosen because of the safety purpose of a study participant in wearing the exoskeleton. The study participant will not get an injury when he carries out experimental tests to determine the optimal values of Kp and Ki. Based on the tests conducted for attaching or not attaching to the study participant, the acquired optimal parameters for Kp and Ki were 200 and 20, respectively. The results of transient performance for the selected Kp and Ki are presented in Figure 11. It provides the transient response for both worn (without an exoskeleton) and not worn (with an exoskeleton) by a study participant for the first and second motor, respectively.

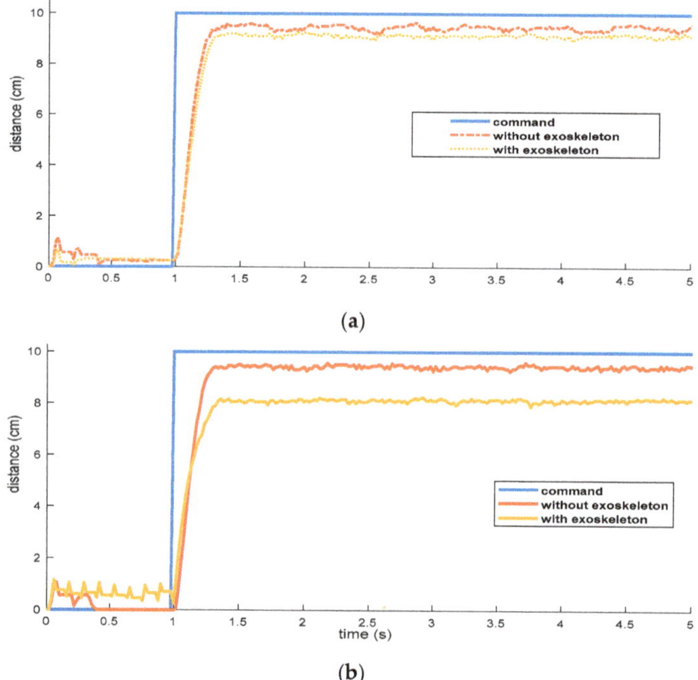


**Figure 11.** Step input response for the (**a**) first motor and (**b**) second motor.

In Figure 11, the command signal is represented by a continuous blue line, the response when it is not worn by a user is depicted by a red graph, and the response when it is worn is illustrated by a yellow graph. Based on the result test in Figure 11, the steady-state response for the second motor is higher than the steady-state error that occurred from the first motor.

The summarized transient performances of the proposed PI control are shown in Tables 2 and 3. In terms of transient response for the exoskeleton when it is worn and not worn, there is a slight difference in the transient performance in both the first and the second motor. However, if it is viewed on the steady-state performance, the occurred steady-state error in exoskeleton which is worn by a study participant has a higher value than that of an exoskeleton which is not worn by a user. The obtained time constant of 1 s is fast enough for the user and it will not harm the user.

**Table 2.** Performance of transient response on the first motor.

| Performance | Symbol | Unattached | Attached | Unit |
| --- | --- | --- | --- | --- |
| Time constant | $\tau$ | 1.095 | 1.106 | s |
| Rise time | $T_r$ | 1.129 | 1.148 | s |
| Settling time | $T_s$ | 4.384 | 4.427 | s |
| Delay time | $T_d$ | 1.072 | 1.079 | s |
| Steady-state error | $E_s$ | 0.21 | 0.95 | cm |

**Table 3.** Performance of transient response on the second motor.

| Performance | Symbol | Unattached | Attached | Unit |
|---|---|---|---|---|
| Time constant | $\tau$ | 1.102 | 1.059 | s |
| Rise time | $T_r$ | 1.135 | 1.093 | s |
| Settling time | $T_s$ | 4.409 | 4.237 | s |
| Delay time | $T_d$ | 1.078 | 1.040 | s |
| Steady-state error | $E_s$ | 0.23 | 1.93 | cm |

In this study, the approached transfer function of the exoskeleton is simulated by giving the step input and then compared to the experimental result. The test results are shown in Figure 12 for both motors. In the figures, the simulated signal is represented by a blue line and a red graph illustrates the experimental signal. The performance for steady-state response in the experimental work has a higher steady-state error than the steady-state error in the simulation based on the transfer function model in Equation (3) through (6). This can occur because of the un-modeled dynamic of the friction force between lead screw and nut, as well as Bowden cable and the exoskeleton.

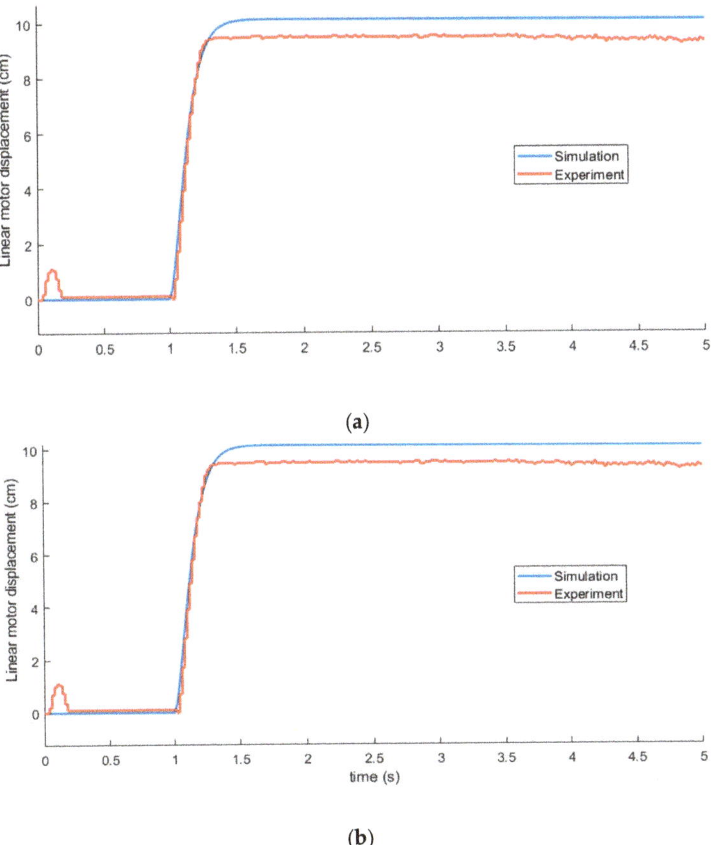


**Figure 12.** The result of simulation vs. experiment during flexion movement: (**a**) First motor; (**b**) second motor.

The obtained Kp and Kd from the previous test are then utilized as the controller for controlling the motion for flexion/extension and pronation/supination. The command for flexion/extension can be achieved by giving the same linear displacement command for both motors, whereas the pronation/supination movement for the exoskeleton is performed by providing the same linear displacement command signal in the opposite direction. The direction of pronation and supination implemented in this study is presented in Figure 13. The first motor is connected to the right side of the wrist hook while the second motor is connected to the left side of the wrist hook. The soft elbow exoskeleton is worn by the user on the left arm as shown in Figure 13.

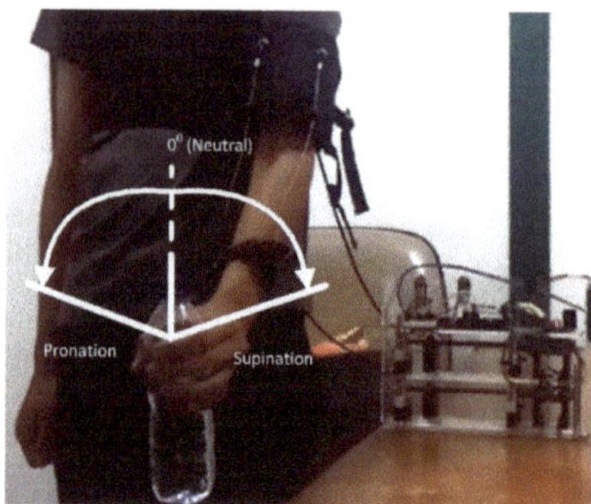


**Figure 13.** Pronation and supination movements in the exoskeleton.

In the first test for pronation/supination assistance, the exoskeleton is given by a commanded signal to provide the mechanical assistance for pronation movement from a neutral position. In this case, the Bowden cable from the second motor will pull the exoskeleton worn by the user and the Bowden cable from the first motor will stretch with the actuator initial position of 6 cm. Because of this mechanism, the resulted steady-state error on the second motor has a higher value than the occurred steady-state error on the first motor as shown in Figure 14.

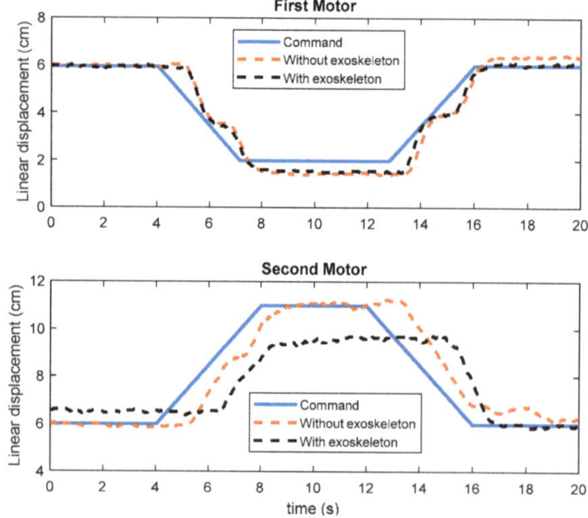


**Figure 14.** Pronation test.

In the second test, the soft exoskeleton is provided by commended signals to move the user arm and wrist from the neutral position to supination motion and hold for five seconds, then go back to the neutral position again. The first motor will pull the wrist and arm, and the second motor will stretch the Bowden cable. In this test, the initial position on the actuator starts at 6 cm. This will contribute to the resulted steady-state error in the first motor as shown in Figure 15.

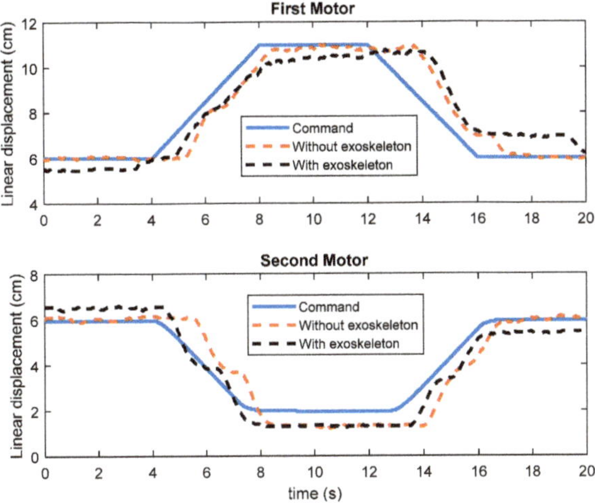

**Figure 15.** Supination test.

The soft exoskeleton is worn to provide mechanical support for combined motion such as flexion, extension, pronation, and supination movements. In this test, the user is grasping bottled water weighing 300 g. The exoskeleton is assigned to provide mechanical force assistance in the following order: The initial position, flexion motion, hold, supination, hold, back to the neutral position, hold, pronation, hold, back to the neutral position, hold, extension, back to initial position. The commanded signal and the response for both motors are presented in Figure 16. The sequence images of the test are documented in Figure 17. Based on the test result, the developed soft elbow exoskeleton can successfully provide mechanical assistance for the user for flexion, extension, pronation, and supination movements. Moreover, the soft exoskeleton is easily attached and detached on the user's arm and wrist.

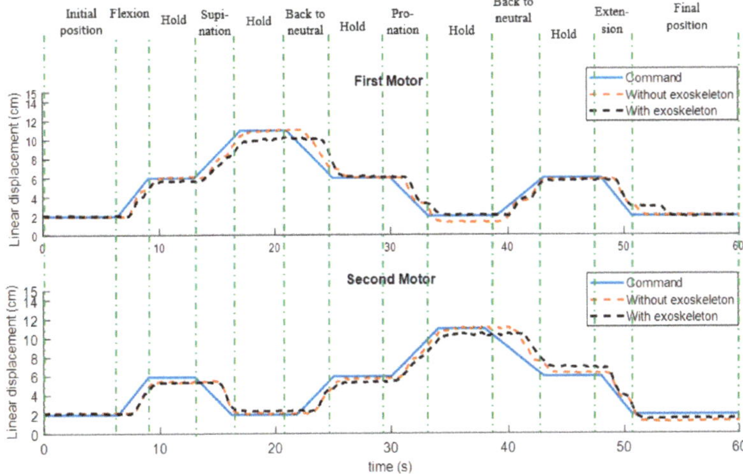

**Figure 16.** Two DOF mechanical assistance for flexion/extension and pronation/supination motion.

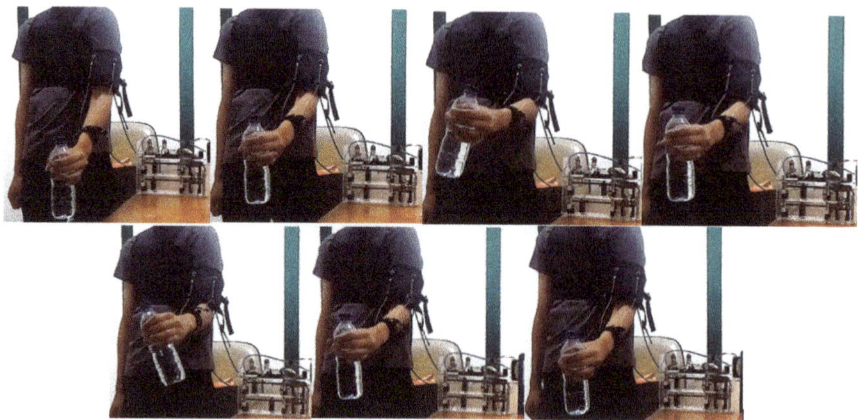

**Figure 17.** Sequence images of two DOF mechanical assistance for flexion/extension and pronation/supination motion.

In order to determine the correlation between the motor-tendon displacement and the resulted angular displacement on the elbow (flexion/extension motion), the motion test is performed by providing the displacement input signal on the motor-tendon actuation system and recording the angular displacement of the elbow. This test is carried out by attaching the soft exoskeleton to the user and then record the changes in the rotational motion of elbow movement that occurs in the soft elbow exoskeleton per unit of distance of the motor-tendon displacement implemented in this study. Table 4 shows the summarized results of the motion tested by acquiring the changes in the angular motion of movement at the elbow.

**Table 4.** Testing for flexion motion assistance.

| No | Obstacle Position | Motor-Tendon Actuator and Soft Elbow Position |
|---|---|---|
| 1. | **Initial/Normal Position** The reading of the IR-Sensor starts from 2 cm which is the starting point (= 0 cm in the starting position). | |
| 2. | **134.91° Position** The IR-Sensor reading is 4 cm in position, moving the arm to an angle of 134.91°. | |

**Table 4.** *Cont.*

| No | Obstacle Position | Motor-Tendon Actuator and Soft Elbow Position |
|---|---|---|
| 3. | **124.58° Position** <br> The IR-Sensor reading is at 6 cm, moving the arm to form an angle of 124.28°. | |
| 4. | **90° Position** <br> The IR-Sensor reading is 9 cm in position, moving the arm in the angle of 90°. | |

The working principle of the proposed motor-tendon actuation system for flexion/extension motion is illustrated in Figure 18. It shows the displacement on 'x' axis coordinate measurements in the soft elbow exoskeleton test. This displacement affects the angular changes in elbow movement which occurs on the 'y' axis shown in Figure 19. The initial position starts from 2 cm which means the zero point (0 cm) testing is at 2 cm because the sensor reading starts at 2 cm and the maximum pulling distance is 12 cm. The initial position of 2 cm is the normal position when the device is attached to the study participant (157.33°) and 9 cm is the final position in the test (90°). In this study, the elbow angle of 90° is sufficient for the flexion/extension motion. This angle is selected based on the consideration of the possible displacement and the safety reason in the actuator although the maximum possible displacement is 12.5 cm.

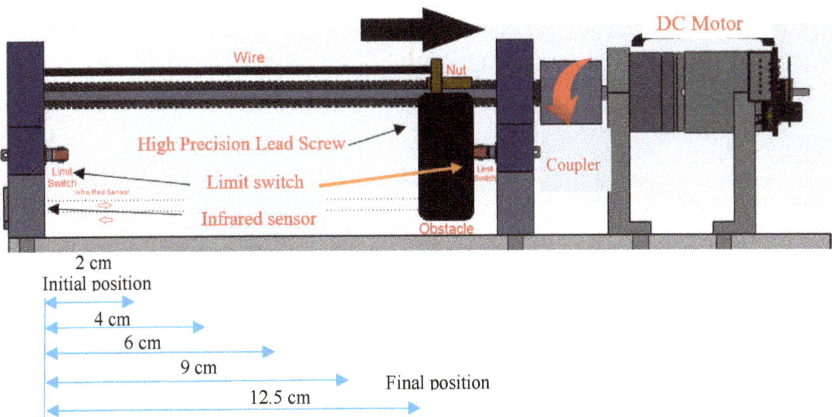


**Figure 18.** Obstacle displacement in motor-tendon actuator.

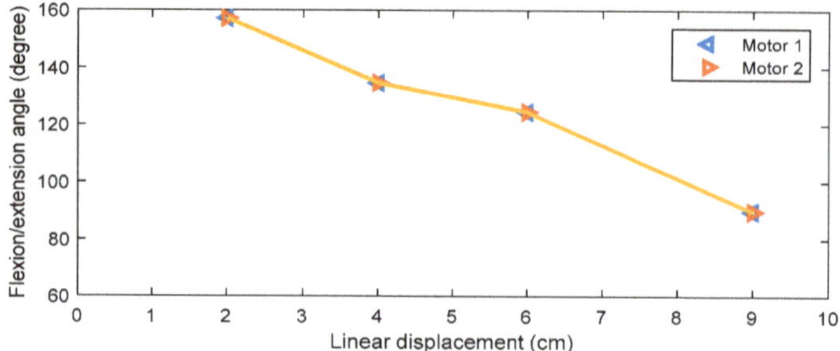

**Figure 19.** Displacement of the motor-tendon actuator vs. elbow output angle for flexion/extension assistance.

The tests carried out on the movements of the angular elbow motion in accordance with Table 4 produces a form of movement in Figure 18. This change is obtained from the displacement of the obstacle distance (cm) to the changes in angular displacement of an elbow in a degree (°). Based on the obtained elbow angle measurement with respect to the linear displacement on the actuator, the soft exoskeleton can provide flexion-extension motion from 157° to 90° as shown in Figure 19. On the other hand, the resulted ROM for flexion/extension motion is 67°. This ROM can be increased by giving the linear displacement command more than 9 cm and less than 12.5 cm.

The resulted angles of the pronation and supination movements are presented in Figure 20. This test is performed without grasping an object. The pronation and supination angles are measured on the user's wrist. Based on the test result, there is similar pattern between the pronation and supination angles as shown in Figure 20. The obtained ROM for the maximum pronation angle is 19° while the acquired maximum supination angle is 18°. These angles are attained when the upper right arm hook is connected to the right wrist hook and the left upper arm hook is connected to the left side of the wrist hook.

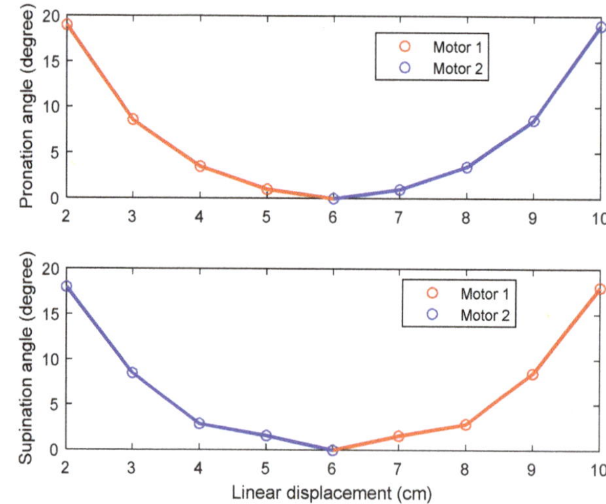

**Figure 20.** Displacement of the motor-tendon actuator vs. output angle for pronation and supination assistance.

## 5. Conclusions

A motor-tendon actuator system for the soft elbow exoskeleton mechanism which is able to assist two DOF elbow and wrist motion is presented. Compared to another actuator mechanism in other previous research, this research employs a motor-tendon actuator which can accommodate elbow and wrist movement in two DOF. Flexion and extension movement for the first DOF, pronation and supination for the second DOF. PI control with a gain value of Kp = 200 and Ki = 20 are applied in both motors resulting in good precision and transient response performance. The output signal from both motors can follow the input with a low value of steady-state error and fast response. When the exoskeleton is worn and not worn on the tests, there is a slight difference in the transient performance on the motor-tendon actuator. The resulted steady-state error in exoskeleton worn by a user has a higher value than that of exoskeleton that is not worn by a study participant. The Bowden cable mechanism applied on the motor-tendon actuator can provide mechanical support for the user's elbow which is needed for a person with muscle function loss in the elbow.

The proposed exoskeleton can provide mechanical assistance for flexion and extension motion ranging from 90° to 157°. The selected maximum ROM for the flexion/extension motion is 67°. This angle can be increased by providing the linear displacement on the actuator more than 9 cm and less than 12.5 cm. The acquired maximum for pronation and supination angles is 19° and 18°, respectively. In the next study, the maximum of pronation and supination angles will be improved by connecting the left upper arm hook to the right wrist hook and the right upper arm hook to the left wrist hook. This can extend the moment arm for pronation-supination motion.

The proposed soft exoskeleton elbow has a lightweight mass that can be worn by the user comfortably. The total mass of the resulted soft elbow exoskeleton including arm and wrist hook is 358 g, while the dual motor-tendon actuator has a mass of 1655 g. The soft exoskeleton can be easily attached and detached by a user. The detailed specification of the proposed soft exoskeleton is summarized in Appendix A. The Simulink block diagram of embedded control and data acquisition block can be found and downloaded in the Supplementary Materials. The PI control block is embedded into Arduino MEGA and the data acquisition block is run on the computer. The 3D design in SolidWorks, engineering drawings, and the STL files, as well as the video of the exoskeleton can be downloaded in our website. It can be found in the Supplementary Materials.

Based on the conducted previous test results, the soft upper arm exoskeleton was successfully developed in providing the mechanical force/support for flexion, extension, pronation, and supination movements. This two DOF movement is relatively easy to control by employing the PI control. The control is commonly widely used for upper arm hard and soft exoskeletons which are mostly used for flexion/extension motion. The performance of the proposed elbow soft exoskeleton for providing mechanical assistance in flexion/extension and pronation/supination tests can be seen online on YouTube at http://y2u.be/M1AKREg3WaY.

In the future, the prototype of the soft exoskeleton will be enhanced in terms of the control input using two channels sEMG signals incorporated machine learning technique such as a neural network. The machine learning will be implemented to discriminate the intended gesture to drive the flexion/extension and pronation/supination movement. A nonlinear control technique will be developed in order to reduce the steady-state error. The size of the actuator system for soft exoskeleton is still big enough to wear. The size of the motor-tendon actuation system will be optimized in order to achieve the optimal size, which is easy to attach on a user's waist. Inertial measuring unit (IMU) which is comprised of accelerometer, gyroscope, and magnetometer will be incorporated on the elbow soft exoskeleton. The IMU will enable the exoskeleton to be controlled directly by flexion/extension and pronation/supination angle command.

**Supplementary Materials:** The Simulink block diagram for controlling the soft exoskeleton utilizing PI compensator, technical drawings, STL files, and 3D design can be downloaded from the following website: https://cbiom3s.undip.ac.id/elbow-exoskeleton/.

**Author Contributions:** R.I. developed the design of the soft elbow exoskeleton, performed the experiment, and wrote the experimental results. M.A. determined the optimized PI control and performed experimental work. I.A.P. and R.A. developed and built the soft elbow exoskeleton using material from fabric. F.T.P. wrote the introduction and provided plots and tables from the experimental work. A.G. provided the suggestion to improve the research such as measurement and experiment as well as the analysis. W.C. proposed and analyzed the result of the measurement and the experimental results. All authors read and approved its content.

**Funding:** This research is funded by Diponegoro University research under Undip Excellence of research (Riset Unggulan Undip) in 2019.

**Conflicts of Interest:** The authors declare no conflict of interest with respect to the research, authorship, and/or publication of this article.

## Appendix A

To provide a clear technical specification of the developed soft exoskeleton, this appendix summarizes the test results that have been carried out and presented in this study.

**Table A1.** The developed soft elbow exoskeleton specification.

| Specification | Value |
| --- | --- |
| Upper arm and wrist hook | Polylactic acid (PLA) |
| Vest material | Polyester |
| Vest mass and hooks | 358 g |
| Dual motor-tendon actuator | 1655 g |
| Dual motor-tendon actuator | $29 \times 9.5 \times 21$ cm |
| Bowden cable length of the first motor and second motor | 70 cm |
| Input voltage | 12 V DC |
| Microcontroller | Arduino MEGA 2560 |
| Maximum of the actuator linear displacement | 12.5 cm |
| Feedback | Infrared sensor |
| Control | Proportional-Integral compensator |
| Feedback | Infrared sensor |
| Motor type | Brushed DC motor |
| Mechanical drive system | Precision lead screw |
| Number of DOF | Two (flexion/extension and pronation/supination) |
| Limit or home sensing | Limit switch |
| Maximum pull force of the first motor | 33 N |
| Maximum pull force of the second motor | 31 N |
| Input sensor as a command | potentiometer |
| Communication interface | USB |
| ROM for flexion/extension | $90°–157°$ |
| ROM for pronation | $0°–19°$ |
| ROM for supination | $0°–18°$ |

## References

1. Chiaradia, D.; Xiloyannis, M.; Solazzi, M.; Masia, L.; Frisoli, A. Comparison of a Soft Exosuit and a Rigid Exoskeleton in an Assistive Task. In *Proceedings of the Wearable Robotics: Challenges and Trends*; Carrozza, M.C., Micera, S., Pons, J.L., Eds.; Springer International Publishing: Cham, Switzerland, 2019; pp. 415–419.

2. Nadas, I.; Pisla, D.; Ceccarelli, M.; Vaida, C.; Gherman, B.; Tucan, P.; Carbone, G. Design of Dual-Arm Exoskeleton for Mirrored Upper Limb Rehabilitation. In *Proceedings of the New Trends in Medical and Service Robotics*; Carbone, G., Ceccarelli, M., Pisla, D., Eds.; Springer International Publishing: Basel, Switzerland, 2019; pp. 303–311.

3. Hein, C.M.; Maroldt, P.A.; Brecht, S.V.; Oezgoecen, H.; Lueth, T.C. Towards an Ergonomic Exoskeleton Structure: Automated Design of Individual Elbow Joints. In Proceedings of the 2018 7th IEEE International Conference on Biomedical Robotics and Biomechatronics (Biorob), Enschede, The Netherlands, 26–29 August 2018; pp. 646–652.

4.  Soltani-Zarrin, R.; Zeiaee, A.; Eib, A.; Langari, R.; Robson, N.; Tafreshi, R. TAMU CLEVERarm: A novel exoskeleton for rehabilitation of upper limb impairments. In Proceedings of the 2017 International Symposium on Wearable Robotics and Rehabilitation (WeRob), Houston, TX, USA, 5–8 November 2017; pp. 1–2.

5.  Tang, Z.; Zhang, K.; Sun, S.; Gao, Z.; Zhang, L.; Yang, Z. An Upper-Limb Power-Assist Exoskeleton Using Proportional Myoelectric Control. *Sensors* **2014**, *14*, 6677–6694. [CrossRef] [PubMed]

6.  Kleinjan, J.G.; Dunning, A.G.; Herder, J.L. Design of a Compact Actuated Compliant Elbow Joint. *Int. J. Struct. Stab. Dyn.* **2014**, *14*, 1440030. [CrossRef]

7.  Vitiello, N.; Lenzi, T.; Roccella, S.; Rossi, S.M.M.D.; Cattin, E.; Giovacchini, F.; Vecchi, F.; Carrozza, M.C. NEUROExos: A Powered Elbow Exoskeleton for Physical Rehabilitation. *IEEE Trans. Robot.* **2013**, *29*, 220–235. [CrossRef]

8.  Pineda-Rico, Z.; Sanchez De Lucio, J.A.; Martinez Lopez, F.J.; Cruz, P. Design of an Exoskeleton for Upper Limb Robot-Assisted Rehabilitation Based on Co-Simulation. Available online: https://www.jvejournals.com/article/16857 (accessed on 27 August 2019).

9.  Lu, L.; Wu, Q.; Chen, X.; Shao, Z.; Chen, B.; Wu, H. Development of a sEMG-based torque estimation control strategy for a soft elbow exoskeleton. *Robot. Auton. Syst.* **2019**, *111*, 88–98. [CrossRef]

10. Wu, W.; Fong, J.; Crocher, V.; Lee, P.V.S.; Oetomo, D.; Tan, Y.; Ackland, D.C. Modulation of shoulder muscle and joint function using a powered upper-limb exoskeleton. *J. Biomech.* **2018**, *72*, 7–16. [CrossRef] [PubMed]

11. Hamaya, M.; Matsubara, T.; Noda, T.; Teramae, T.; Morimoto, J. Learning assistive strategies for exoskeleton robots from user-robot physical interaction. *Pattern Recognit. Lett.* **2017**, *99*, 67–76. [CrossRef]

12. Ganesan, Y.; Gobee, S.; Durairajah, V. Development of an Upper Limb Exoskeleton for Rehabilitation with Feedback from EMG and IMU Sensor. *Procedia Comput. Sci.* **2015**, *76*, 53–59. [CrossRef]

13. Koh, T.H.; Cheng, N.; Yap, H.K.; Yeow, C.-H. Design of a Soft Robotic Elbow Sleeve with Passive and Intent-Controlled Actuation. *Front. Neurosci.* **2017**, *11*, 597. [CrossRef] [PubMed]

14. Zhu, Y.; Zhang, G.; Li, H.; Zhao, J. Automatic load-adapting passive upper limb exoskeleton. *Adv. Mech. Eng.* **2017**, *9*, 1687814017729949. [CrossRef]

15. Copaci, D.; Cano, E.; Moreno, L.; Blanco, D. New Design of a Soft Robotics Wearable Elbow Exoskeleton Based on Shape Memory Alloy Wire Actuators. *Appl. Bionics Biomech.* **2017**, *2017*, 1605101. [CrossRef] [PubMed]

16. D'Angeles Mendes De Brito, A.C.; Kutilek, P.; Hejda, J.; Smrcka, P.; Havlas, V. Design of Smart Orthosis of Upper Limb for Rehabilitation. In *Proceedings of the World Congress on Medical Physics and Biomedical Engineering 2018*; Lhotska, L., Sukupova, L., Lacković, I., Ibbott, G.S., Eds.; Springer: Singapore, 2019; pp. 773–778.

17. Wei, W.; Qu, Z.; Wang, W.; Zhang, P.; Hao, F. Design on the Bowden Cable-Driven Upper Limb Soft Exoskeleton. In *Proceedings of the Applied Bionics and Biomechanics*; Hindawi: London, UK, 2018.

18. Qingcong, W.; Xingsong, W. Design of a Gravity Balanced Upper Limb Exoskeleton with Bowden Cable Actuators. *IFAC Proc. Vol.* **2013**, *46*, 678–683. [CrossRef]

19. Marconi, D.; Baldoni, A.; McKinney, Z.; Cempini, M.; Crea, S.; Vitiello, N. A novel hand exoskeleton with series elastic actuation for modulated torque transfer. *Mechatronics* **2019**, *61*, 69–82. [CrossRef]

20. In, H.; Kang, B.B.; Sin, M.; Cho, K. Exo-Glove: A Wearable Robot for the Hand with a Soft Tendon Routing System. *IEEE Robot. Autom. Mag.* **2015**, *22*, 97–105. [CrossRef]

21. Manna, S.K.; Dubey, V.N. Comparative study of actuation systems for portable upper limb exoskeletons. *Med. Eng. Phys.* **2018**, *60*, 1–13. [CrossRef] [PubMed]

22. Kong, K. Proxy-based impedance control of a cable-driven assistive system. *Mechatronics* **2013**, *23*, 147–153. [CrossRef]

23. Noda, T.; Teramae, T.; Ugurlu, B.; Morimoto, J. Development of an upper limb exoskeleton powered via pneumatic electric hybrid actuators with bowden cable. In Proceedings of the 2014 IEEE/RSJ International Conference on Intelligent Robots and Systems, Chicago, IL, USA, 14–18 September 2014; pp. 3573–3578.

24. Aguilar-Sierra, H.; Yu, W.; Salazar, S.; Lopez, R. Design and control of hybrid actuation lower limb exoskeleton. *Adv. Mech. Eng.* **2015**, *7*, 1687814015590988. [CrossRef]

25. Gao, X.; Sun, Y.; Hao, L.; Yang, H.; Chen, Y.; Xiang, C. A New Soft Pneumatic Elbow Pad for Joint Assistance With Application to Smart Campus. *IEEE Access* **2018**, *6*, 38967–38976. [CrossRef]

26. Xiloyannis, M.; Chiaradia, D.; Frisoli, A.; Masia, L. Physiological and kinematic effects of a soft exosuit on arm movements. *J. NeuroEng. Rehabil.* **2019**, *16*, 29. [CrossRef] [PubMed]

 *electronics*

*Review*

# Insight on Electronic Travel Aids for Visually Impaired People: A Review on the Electromagnetic Technology

**Emanuele Cardillo * and Alina Caddemi**

Department of Engineering, University of Messina, Messina 98166, Italy; acaddemi@unime.it
* Correspondence: ecardillo@unime.it

Received: 15 October 2019; Accepted: 28 October 2019; Published: 4 November 2019

**Abstract:** This review deals with a comprehensive description of the available electromagnetic travel aids for visually impaired and blind people. This challenging task is considered as an outstanding research area due to the rapid growth in the number of people with visual impairments. For decades, different technologies have been employed for solving the crucial challenge of improving the mobility of visually impaired people, but a suitable solution has not yet been developed. Focusing this contribution on the electromagnetic technology, the state-of-the-art of available solutions is demonstrated. Electronic travel aids based on electromagnetic technology have been identified as an emerging technology due to their high level of achievable performance in terms of accuracy, flexibility, lightness, and cost-effectiveness.

**Keywords:** electronic travel aids; radar; electromagnetic; visual impairment; blind people; microwaves; mobility; antennas

---

## 1. Introduction

It is estimated that 188.5 million people worldwide have mild vision impairment, 217 million have moderate-to-severe vision impairment, and 36 million people are blind [1].

This is a continuously growing number that has fueled the interest of many in in the pursuit of improving the quality of life of those affected by blindness [2–6].

Usually, the mobility of visually impaired people is addressed by the use of travel supports, with the traditional white-cane being the most renowned and still the most popular. However, there are many electronic travel aids (ETAs) available to make mobility safer and more familiar for both inside and outside use.

The National Research Council published a list of the most important requirements of an ETA [7,8]:

1) Detection of obstacles from ground level to height of the head;
2) Information about the travel surface;
3) Detection of objects bordering the travel path for shore-lining and projection;
4) Distant object and cardinal direction information for the projection of a straight line;
5) Location of landmarks and identification information;
6) Information for enhancing self-familiarization and for creating a mental map of the environment.

These six points represent the landmark for designing an effective ETA. The first three requirements provide indications about the obstacles that an ETA should be able to detect. Points four to six suggest additional information is needed for easier and safer movement in a complex environment.

Since eighty years ago, when the first ETAs were introduced, the work of researchers has concentrated on the design of better aids, but as yet these have not materialized. [9]. To date, none of the available aids has fulfilled all the requirements to fully satisfy users' expectations.

The ETAs are usually divided into two categories: primary and secondary aids. Primary aids allow autonomous movement without other aids since they provide all the required information independently. Secondary aids only enhance the mobility and must be used in conjunction with other aids (e.g., primary aids). Moreover, ETAs can be classified as mobility and navigation aids, where the former help the user in detecting and avoiding obstacles in their path and the latter allow to the user to find the path itself from the starting point to the destination [10].

Different technologies have been used with the aim of creating better devices, the most significant being based on ultrasonic sensors and optical/vision systems [11–17].

Although technological advancement has reached an elevated level of maturity and a variety of devices are available on the market, ultrasonic systems have limitations [10]. To specify, the maximum detectable range is limited when smooth surfaces need to be detected, even worse if the angle of incidence of the ultrasonic beam is small. Furthermore, due to the relatively large radiation beam, it is difficult for small apertures to be accurately detected (e.g., the presence of doors). Nevertheless, this kind of aid is widely used due to the easy-to-access technology and the availability of low-cost ultrasonic sensors.

ETAs based on optical technologies are very robust in terms of object detection capabilities under normal operating conditions. They typically employ cameras, and the target recognition is provided by means of complex algorithms [18]. The main drawbacks are due to the high sensitivity to ambient light, which can seriously limit the system performance, especially in outdoor environments [10]. Additional issues are the strong dependence on the optical properties of obstacles and the bulky dimensions of the devices.

Due to the lack of a competitive and robust technology, ETA devices based on the electromagnetic technology have only been developed in the past decade; These devices usually achieve target detection by sending electromagnetic waves into the space and analyzing the reflected signal.

Most aids are radar-based systems, where the frequency-modulated continuous wave (FMCW) is the typical mode of operation [19–25]. In a FMCW radar, the information about the target is contained in the frequency of the received signal. The transmitted signal is modulated (e.g., using a linear frequency modulation or phase modulation), reflected back by the target, and conditioned by a homodyne receiver. Thereafter, it is possible to extract the range information from the intermediate frequency (IF) signal. For example, by considering a saw-tooth modulation, the range ($R$) of the target can be computed by using (1):

$$R = \frac{c f_{IF} \tau}{2B},\tag{1}$$

where $B$ and $\tau$ are the modulation bandwidth and the duration of the signal, respectively, $c$ is the speed of light, and $f_{IF}$ is the frequency of the IF signals whose amplitudes overcome a certain threshold [26,27].

The advantages of this technology compared to ultrasonic and optical systems can be summarized as follows:

- Since higher frequency corresponds to a smaller wavelength, at frequencies typical of microwave and millimeter waves, the circuits become very compact and lightweight [28,29].
- The larger the modulation bandwidth of the transmitted signal, the better the resolution [28,29]. This feature allows, for example, for two near objects to be distinguished from each other.
- The radiation pattern can be tailored by modifying the design of the antennas, thus accurately detecting small apertures or suspended obstacles.

To the best knowledge of the authors, this is the first review article concerning ETAs based on electromagnetic technology. This review is necessary due to the previous lack of a paper that collects and illustrates the main results in this field.

In Section 2, the pertinent studies and prototypes based on the electromagnetic technology are reported. Particular attention is given to the illustration of both practical and technological aspects by clearly explaining the basic modes of operation of the considered aids.

In Section 3, the particular differences between the devices are analyzed providing, in the authors' opinion, the possible future developments of this technology. Finally, the conclusions are outlined in Section 4.

## 2. Electromagnetic Aids for Visually Impaired and Blind People

The idea of employing an electromagnetic system as an aid for visually impaired people is quite recent. In [30], a vector network analyzer and a double-ridge horn antenna were used for transmitting and receiving a sequence of pulses from 1 to 6 GHz. Then, an inverse Fourier transform was applied to the frequency-domain measurements in order to recreate a very short time-domain pulse. The measured delay of the received pulse corresponds to the round-trip time-of-flight (TOF) of the electromagnetic wave from the transmitting antenna to the obstacle. Thereafter, the range of the object could be easily computed [31]. The diagram of the system and the echo pulse reflected by a $25 \times 25$ cm$^2$ metal plate at a distance of 2 m are shown in Figure 1.

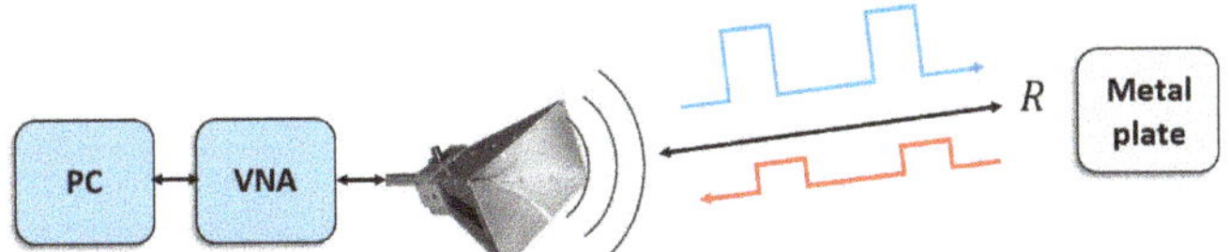


**Figure 1.** Diagram of the system.

The proposed approach was compared with an ultrasonic system, making the two experimental set-ups as equivalent as possible. The proposed system showed better precision and the ability to detect obstacles that are not detectable by ultrasonic systems (e.g., a partially open door). Although the system is very bulky and expensive, its future miniaturization has been claimed.

An enhanced version of the system proposed in [30] has been reported in [32]. A portable vector network analyzer was employed for testing the effectiveness of the methodology in both indoor and outdoor scenarios. The maximum range was fixed to 3 m, whereas the operating frequencies from 4 to 6 GHz achieved a spatial resolution of 12 cm. The tests were carried out by a blind volunteer during different test conditions (i.e., obstacles located in different arrangements). The test set-up was arranged in order to have the obstacles aligned, shaped as a cage, and with an obstacle at chest level—as depicted in Figure 2. The system could correctly detect the presence of one or more obstacles even at the level of the head or chest (e.g., an open window or a tree branch).

(a)

**Figure 2.** *Cont.*

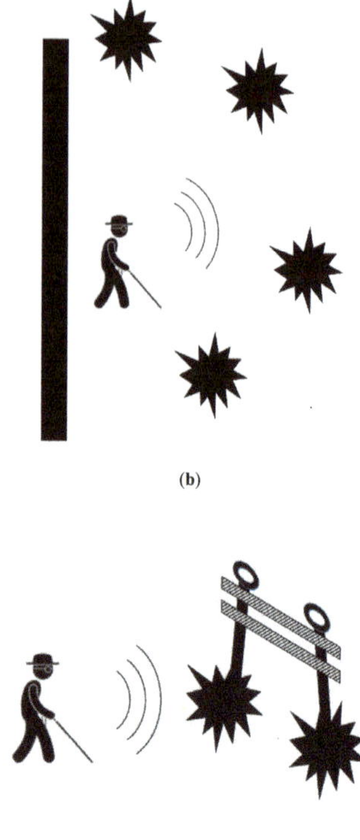

(b)

(c)

**Figure 2.** Test set-up with (**a**) aligned obstacles, (**b**) obstacles shaped as a cage, and (**c**) obstacles at chest level.

A peculiar application of an electromagnetic aid for visually impaired people is reported in [33], where a system capable of guiding a blind athlete during running and physical activities was designed, realized, and tested. This arose from the need to enhance the runners' freedom during competitions or training. As a matter of fact, athletes currently need to be guided by a non-extendible cable carried by another runner, thus limiting the possibility of free movement [34].

The proposed system is composed of: (a) a mobile unit preceding the runner that generates two "electromagnetic walls" (i.e., two electromagnetic waves radiated on the left and right sides of the athlete) and (b) a receiving unit worn by the runner equipped with vibro-tactile devices.

A representation of the guiding system is shown in Figure 3.

**Figure 3.** Representation of the guiding system [33].

The mobile unit generates a vibro-tactile warning when the athlete is leaving the area delineated by the electromagnetic walls, to guide them towards the center of the path.

In this circumstance, the use of electromagnetic waves ensures confinement of the wall to a narrow and well-defined area. The operating bandwidth of the system is inside the X-band, thus reducing the dimension and weight of the antenna in the receiving unit worn by the blind runner. To achieve this, the antenna was designed with lightweight and compact microstrip technology. The system was tested by a blind athlete using four different paths as reported in Figure 4, demonstrating the proper operation of the prototype. A video of the tests can also be watched online [34].

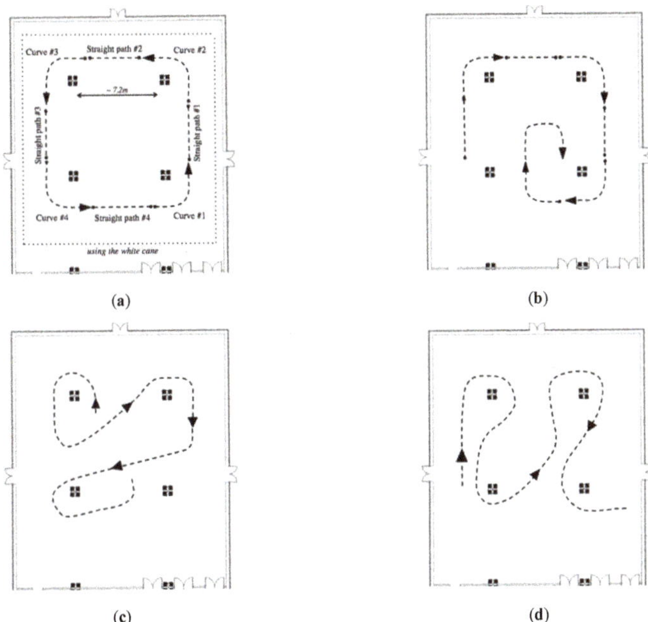


**Figure 4.** Paths traveled by the athlete in a rectangular room ($20 \times 17$ m$^2$): (**a**) square, (**b**) spiral, and (**c**,**d**) complex paths [33].

In [19], the challenge of avoiding collision with obstacles for visually impaired users was achieved by creating a radar-based navigation device. The system employs a frequency-modulated continuous-wave signal. Although the authors plan to have a main sensor working at the center frequency of 80 GHz with a bandwidth of 24 GHz, a bandwidth of 4 GHz—from 80 GHz to 84 GHz—was adopted for the tests. With this frequency bandwidth, a spatial resolution equal to 6 cm can be obtained. The radar sensor was attached on the head of the blind person, whose position was continuously tracked by an inertial sensor. This sensor scanned the surrounding environment and detected the position of obstacles up to 5 m. The intention of the authors is to scan the scenario using a beam-steering antenna. Its design is shown in [19], but since it is still under development, the rotation was mechanically achieved by a stepper motor. Due to the size of the setup, the radar was placed on a large mounting. A virtual map of the environment was created and mapped into a 3D audio signal. The authors assert that a long training process is not required.

A superior system performance compared to camera-based systems is claimed in terms of independence from light conditions, ability to detect transparent obstacles such as glass doors, and better visual range in rain or fog [35,36].

The challenge of tailoring a suitable electromagnetic aid is shown in [20], highlighting the effectiveness of the authors' solution. A clinical investigation was performed with 25 visually impaired people wearing a radar-based assistive device.

The working frequency of the system was 24 GHz, selected to facilitate a lightweight wearable device, because at this frequency the signal can penetrate textiles, making it possible to wear the device under clothing. The system automatically compensates small tilt errors or notifies the person if the incline is excessive. It is able to detect obstacles up to 3.5 m, within a horizontal sector of 25° and a vertical sector of 70°.

The maximum usage time of the prototype was measured as between 4 and 5 h, but this can be improved in a future designs. The warning about the presence of obstacles is fed back to the user by means of acoustic or haptic feedback. In Figure 5, the operating principle of the system and a picture of the device is shown.

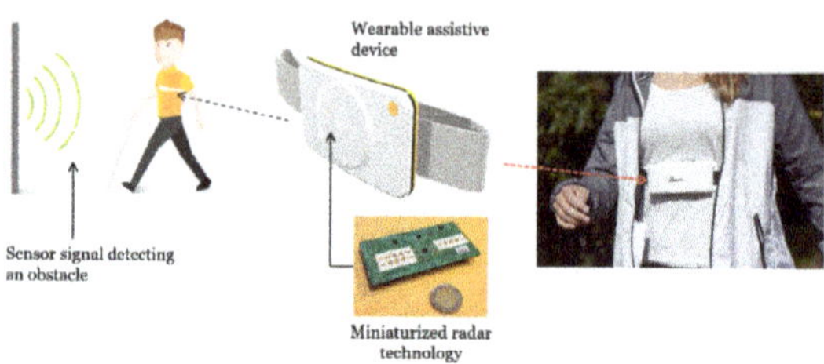


**Figure 5.** Operating principle of the system and picture of the device [20].

The clinical investigation was implemented during an observation period of two weeks, using the electromagnetic aid during everyday activities.

The results of an interview after usage of the device are shown in Figure 6a in response to the question "Does the assistive device help you to perceive your environment better?", whereas in Figure 6b the question was "Does the assistive device increase your confidence in independent walking?".

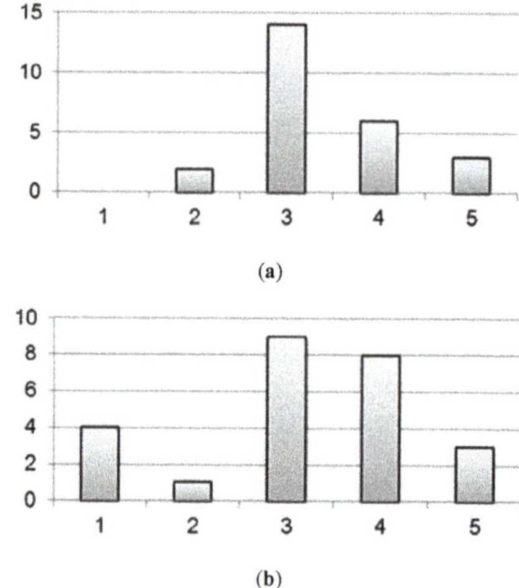


**Figure 6.** Results of the interview questions: (**a**) "Does the assistive device help you to perceive your environment better?" and (**b**) "Does the assistive device increase your confidence in independent walking?" The answers reported on the X-axes are: (1) not at all, (2) not much, (3) to some extent, (4) significantly, (5) very significantly [20].

For both graphs in Figure 6, the answers recorded on the X-axes are: (1) not at all, (2) not much, (3) to some extent, (4) significantly, (5) very significantly.

In the results, 92% of the participants stated that the perception of the surrounding environment was enhanced by the electromagnetic aid, whereas 80% noticed an increased confidence in independent mobility.

Solutions employing 24 GHz radars based on FMCW modulation are presented in [21–23], but different choices were made concerning the way the users employ the devices.

Despite the extensive range of available ETAs, the most widespread aid is still the traditional white-cane, due to its ease of use and cost effectiveness [37,38].

Therefore, the authors decided to create a device so small and lightweight that it can be integrated into a white-cane—a kind of "microwave cane". The authors based their design on feedback from users during interviews and the remarks of visually impaired and blind users, considering the positive reaction to a new device as the main task. Volunteers testing ETAs usually suggest the simultaneous use the white cane.

Since the self-confidence that the traditional white cane is reported to give users is absent in ETAs, the authors chose to go down this route.

Of course, this new device must overcome the limitations of the traditional cane (i.e., it does not provide any protection against obstacles at head or chest height and the range is confined to the length of the cane). At the same time, the system has to be cost effective and easy to use without extensive training.

In [21], great attention is paid to the design of a customized antenna. It was designed by employing microstrip technology in order to have a compact system that can be attached to the cane. They highlight the advantages of the system compared to ones based on ultrasonic or optical technology, overcoming the limitations of limited range and poor performance operating within a low incidence angle on smooth surfaces. The system was tested up to 5 m, with a spatial resolution of 60 cm (24.0–24.25 GHz).

In [22], a feasibility study was undertaken with the aim of accurately identifying the technical specification of the microwave cane. The behavior of the entire system was simulated using computer-aided design (CAD), with preliminary results obtained from newly designed transmitting and receiving antennas. The first experimental activity using the microwave cane is outlined in [23].

The authors obtained system spatial resolution close to 14 cm, reduced energy consumption, a maximum range of 5 m, and horizontal and vertical antenna patterns close to 12° and 40°, respectively.

A picture of the prototype is shown in Figure 7.

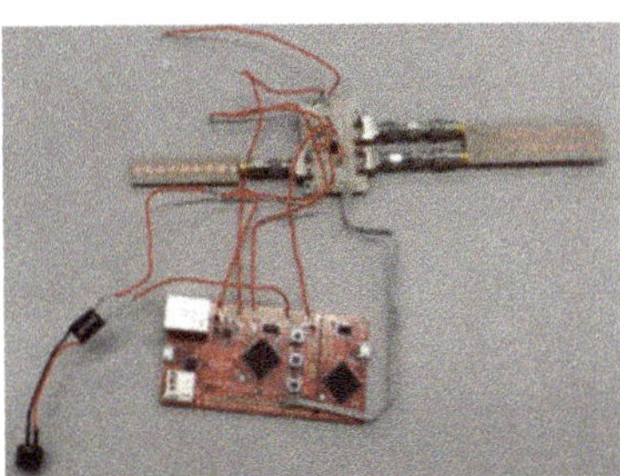

**Figure 7.** Picture of the radar prototype [23].

The radar system was tested with different obstacles commonly used for daily tasks, that is, a wooden chair, a chest of drawers, and a human subject. The targets, characterized by different materials, dimensions, and shapes, were located at the range of 1.5 m and successfully detected, as shown in Figure 8.

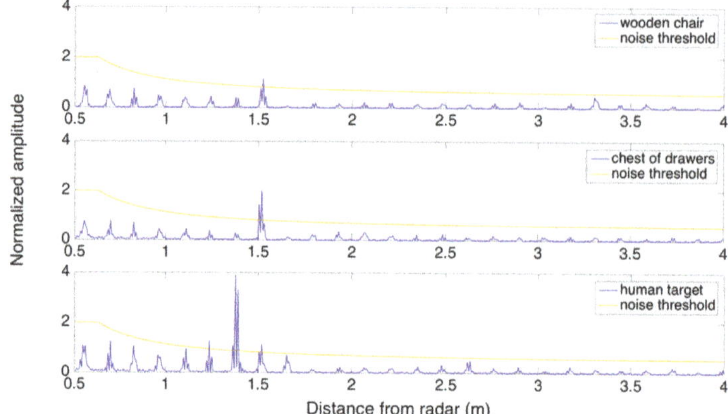

**Figure 8.** Intermediate frequency (IF) signals obtained from obstacles at the range of 1.5 m. From the top to bottom, results are shown for a wooden chair, a chest of drawers, and a human target [23].

An interesting solution is illustrated in [24,25], where the radar technology was employed in conjunction with an RGB-D sensor. The radar operated from 77 to 81 GHz, implementing an FMCW signal. This 4 GHz bandwidth gives a spatial resolution of 6 cm. The purpose of the exercise was to enhance the features of object detection and localization. Whereas mm-wave radar is capable of providing speed information and a relatively high spatial resolution in the radial direction, the RGB-D sensor accomplishes the task of distinguishing two targets at the same distance. A block diagram of the system is shown in Figure 9.

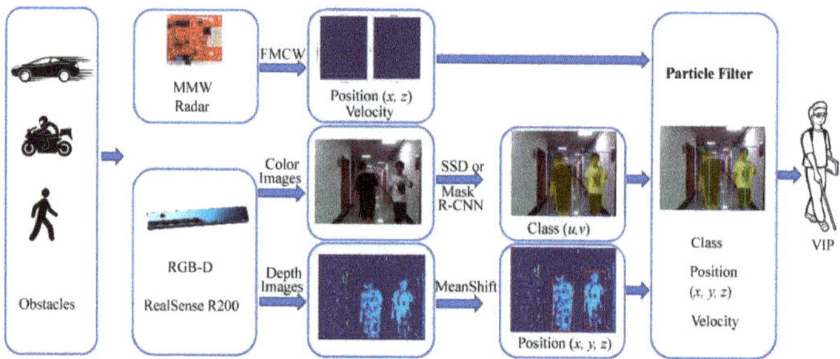


**Figure 9.** Schematic diagram of the system in [24]. FMCW: frequency-modulated continuous wave.

Obstacles at the same distance are recognized using color information and the application of a single deep neural network or object instance segmentation [39,40].

In Figure 10, results relating to tests performed in an indoor environment are shown. In Figure 10(a1,a2) the mask R-CNN (Region convolutional neural network) was used for detecting and recognizing the targets, whereas in (b1,b2) the SSD (Single Shot MultiBox Detector ) network was used. In (c1,c2) the image results are shown with, detection results enclosed in a red box, whereas in (d1,d2) the data detected by the radar are illustrated. Finally, in (e1,e2), the data are fused using a particle filter.

The potential of this technology is also highlighted by the Integrated Smart Spatial Exploration System (INSPEX H2020 project), where the main goal is to take advantage of the automotive techniques concerning spatial exploration and obstacle detection in the design of portable sensors [41]. Potential applications employing radar-based sensors range from safer human navigation in reduced visibility to navigation for visually impaired people. The authors intend to integrate the INSPEX system into a smart white cane.

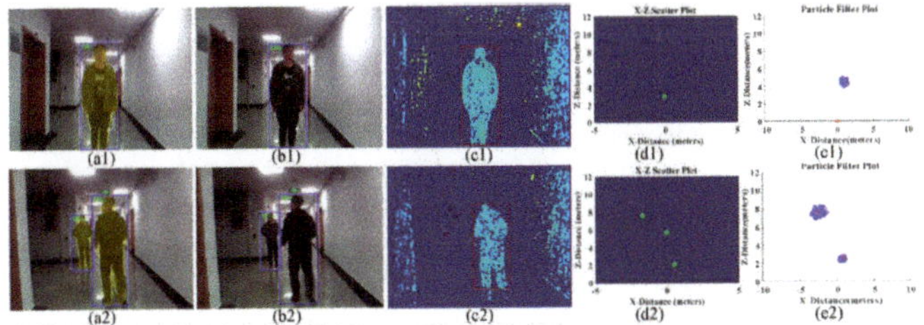

**Figure 10.** Tests performed in an indoor environment and related results [24]. (**a1,a2**) Detection performed by R-CNN, (**b1,b2**) detection performed by SSD network, (**c1,c2**) image results, (**d1,d2**) data detected by the radar, (**e1,e2**), data fused using a particle filter.

## 3. Discussion

In Section 2, the main contributions in the field were reported, showing properties of the proposed systems such as working principle, maximum range, cost, size, spatial resolution, operating frequency, and the existence of a final prototype. These aspects are organized in Table 1.

**Table 1.** Properties of the main electromagnetic aids for visually impaired people.

| Ref. | Working Principle | Maximum Range (m) | Cost | Size | Spatial Resolution (cm) | Operating Frequency (GHz) | Final Prototype |
| --- | --- | --- | --- | --- | --- | --- | --- |
| [32] | VNA-based | 3 | High | Bulky | 12 | 4.0–6.0 | No |
| [19] | Radar | 5 | Medium | Bulky | 6 | 80.0–84.0 | No |
| [20] | Radar | 3.5 | Low | Small | ND | 24.0 | No |
| [21] | Radar | 5 | Low | Small | 60 | 24.0–24.25 | No |
| [23] | Radar | 5 | Low | Small | 14 | 24.0–25.1 | No |
| [24] | Radar + RGB | 12 | Low | Medium | 6 | 77.0–81.0 | No |

By analyzing the properties shown in Table 1, we can conclude that different devices could be market ready very quickly, which would provide useful products that could offer a good compromise between the capability of scanning the environment, cost-effectiveness, and size.

Therefore, it is legitimate to ask why there is still no valid electromagnetic aid available on the market. We believe that it can be explained by the fact that there is insufficient collaboration between research groups and industry, which would provide a more effective transition from laboratory prototypes to the final engineered product. Indeed, since the core technology is based on radar, the necessary components, used by the relevant advancement in the automotive field, are readily available for use Moreover, the miniaturization of the operating frequencies in a typical of microwave and millimeter waves could be utilized to minimize size and weight, affording easy integration in everyday activities.

Although the electromagnetic aids exhibit better performance than ultrasonic or camera-based solutions, recent years have seen these technologies more broadly diffused due to their wider availability in the market. Moreover, in published papers only limited attention has been given to finding the best solution for providing feedback to the user. A greater effort in this direction could be very beneficial for enhancing the usability of this category of travel aid.

## 4. Conclusions

In this review, the most recent developments relating to electromagnetic travel aids for visually impaired and blind people are presented. The main contributions in the field were analyzed, highlighting the working principle, the impact on the end user's life, and the main properties. The presented solutions mostly employ radar-based devices, with only one system exploiting a (vector network analyzer) VNA-based technology. Overall, the trend is towards employing ever-higher operation frequencies in order to realize compact systems and to increase the resolution capabilities. The electromagnetic technology shows better performance than other types of electronic travel aid, as they have a more reliable signal processing technique, smaller dimensions, higher precision and resolution, and better immunity to noise. Although the "perfect" aid has not yet been realized, this contribution highlights that the technology is mature and different solutions can be adopted for fulfilling this challenging task.

**Author Contributions:** E.C. collected the required bibliography and wrote the review article. A.C. supervised the research activities and revised the paper.

**Funding:** This research received no external funding.

**Conflicts of Interest:** The authors declare no conflict of interest.

## References

1. Bourne, R.R.A.; Flaxman, S.R.; Braithwaite, T.; Cicinelli, M.V.; Das, A.; Jonas, J.B.; Keeffe, J.; Kempen, J.H.; Leasher, J.; Limburg, H.; et al. Magnitude, temporal trends, and projections of the global prevalence of blindness and distance and near vision impairment: A systematic review and meta–analysis. *Lancet Glob. Health* **2017**, *5*, 888–897. [CrossRef]
2. Corcoran, C.; Douglas, G.; Pavey, S.; Fielding, A.; McCall, S.; McLinden, M. Network 1000: The changing needs and circumstances of visually impaired people: A project overview. *Br. J. Vis. Impair.* **2004**, *22*, 93–100. [CrossRef]
3. Velazquez, R. Wearable assistive devices for the blind. In *Wearable and Autonomous Biomedical Devices and Systems for Smart Environment*; Lecture Notes in Electrical Engineering; Springer International Publishing: New York, NY, USA, 2010; Volume 75, pp. 331–349.
4. Paiva, S. *Technological Trends in Improved Mobility of the Visually Impaired*, 1st ed.; Springer International Publishing: New York, NY, USA, 2020.
5. Islam, M.M.; Sadi, M.S.; Zamli, K.Z.; Ahmed, M.M. Developing walking assistants for visually impaired people: A review. *IEEE Sens. J.* **2019**, *19*, 2814–2828. [CrossRef]
6. Bai, J.; Liu, Z.; Lin, Y.; Li, Y.; Lian, S.; Liu, D. Wearable travel aid for environment perception and navigation of visually impaired people. *Electronics* **2019**, *8*, 697. [CrossRef]
7. Blasch, B.B.; Wiener, W.R.; Welsh, R.L. *Foundations of Orientation and Mobility*, 3nd ed.; AFB Press: New York, NY, USA, 2010.
8. National Research Council. *Electronic Travel Aids: New Directions for Research*; Nat. Acad. Press: Washington, DC, USA, 1986.
9. Roentgen, U.R.; Gelderblom, G.J.; Soede, M.; De Witte, L. Inventory of electronic mobility aids for visually impaired persons—A literature review. *J. Vis. Impair. Blind.* **2008**, *102*, 702–724. [CrossRef]
10. Di Mattia, V.; Scalise, L.; Petrini, V.; Russo, P.; De Leo, A.; Pallotta, E.; Mancini, A.; Zingaretti, P.; Cerri, G. Electromagnetic technology for a new class of electronic travel aids supporting the autonomous mobility of visually impaired people. *Sensors* **2017**, *17*, 364.
11. Patil, K.; Jawadwala, Q.; Shu, F.C. Design and construction of electronic aid for visually impaired people. *IEEE Trans. Hum. Mach. Syst.* **2018**, *48*, 172–182. [CrossRef]
12. Zhou, D.; Yang, Y.; Yan, H. A smart 'virtual eye' mobile system for the visually impaired. *IEEE Potentials* **2016**, *35*, 13–20. [CrossRef]
13. Sohl–Dickstein, J.; Teng, S.; Gaub, B.M.; Rodgers, C.C.; Li, C.; DeWeese, M.R.; Harper, N.S. A device for human ultrasonic echolocation. *IEEE Trans. Biomed. Eng.* **2015**, *62*, 1526–1534. [CrossRef]
14. Bhatlawande, S.; Mahadevappa, M.; Mukherjee, J.; Biswas, M.; Das, D.; Gupta, S. Design, development, and clinical evaluation of the electronic mobility cane for vision rehabilitation. *IEEE Trans. Neural Syst. Rehabil. Eng.* **2014**, *22*, 1148–1159. [CrossRef]
15. Bai, J.; Lian, S.; Liu, Z.; Wang, K.; Liu, D. Virtual–blind–road following–based wearable navigation device for blind people. *IEEE Trans. Consum. Electron.* **2018**, *64*, 136–143. [CrossRef]
16. Kang, M.-C.; Chae, S.-H.; Sun, J.-Y.; Lee, S.-H.; Ko, S.-J. An enhanced obstacle avoidance method for the visually impaired using deformable grid. *IEEE Trans. Consum. Electron.* **2017**, *63*, 169–177. [CrossRef]

17. Yang, K.; Wang, K.; Hu, W.; Bai, J. Expanding the detection of traversable area with realsense for the visually impaired. *Sensors* **2016**, *10*, 1954. [CrossRef] [PubMed]
18. Cardillo, E.; Caddemi, A. Feasibility study to preserve the health of an Industry 4.0 worker: A Radar System for Monitoring the Sitting-Time. In Proceedings of the IEEE International Workshop on Metrology for Industry 4.0 and IoT, Naples, Italy, 4–6 June 2019.
19. Kwiatkowski, P.; Jaeschke, T.; Starke, D.; Piotrowsky, L.; Deis, H.; Pohl, N. A Concept Study for a Radar-Based Navigation Device with Sector Scan Antenna for Visually Impaired People. In Proceedings of the IEEE MTT–S International Microwave Bio Conference (IMBIOC), Gothenburg, Sweden, 15–17 May 2017.
20. Kiuru, T.; Metso, M.; Utriainen, M.; Metsavainio, K.; Jauhonen, H.M.; Rajala, R.; Savenius, R.; Strom, M.; Jylha, T.N.; Juntunen, R.; et al. Assistive device for orientation and mobility of the visually impaired based on millimeter wave radar technology—Clinical investigation results. *Cogent Eng.* **2018**, *5*, 1–12. [CrossRef]
21. Pisa, S.; Pittella, E.; Piuzzi, E. Serial patch array antenna for an FMCW radar housed in a white cane. *Int. J. Antennas Propag.* **2016**, *2016*, 9458609. [CrossRef]
22. Di Mattia, V.; Manfredi, G.; De Leo, A.; Russo, P.; Scalise, L.; Cerri, G.; Caddemi, A.; Cardillo, E. A Feasibility Study of a Compact Radar System for Autonomous Walking of Blind People. In Proceedings of the International Forum on Research and Technologies for Society and Industry Leveraging a Better Tomorrow (RTSI), Bologna, Italy, 7–9 September 2016.
23. Cardillo, E.; Di Mattia, V.; Manfredi, G.; Russo, P.; De Leo, A.; Caddemi, A.; Cerri, G. An electromagnetic sensor prototype to assist visually impaired and blind people in autonomous walking. *IEEE Sens. J.* **2018**, *18*, 2568–2576. [CrossRef]
24. Long, N.; Wang, K.; Cheng, R.; Hu, W.; Yang, K. Unifying obstacle detection, recognition, and fusion based on millimeter wave radar and RGB–depth sensors for the visually impaired. *Rev. Sci. Instrum.* **2019**, *90*, 1–12. [CrossRef]
25. Long, N.; Wang, K.; Cheng, R.; Yang, K.; Hu, W.; Bai, J. Assisting the visually impaired: Multitarget warning through millimeter wave radar and RGB–depth sensors. *J. Electron. Imag.* **2019**, *28*, 1–15. [CrossRef]
26. Cardillo, E.; Caddemi, A. A novel approach for crosstalk minimization in FMCW radars. *Electron. Lett.* **2017**, *53*, 1379–1381. [CrossRef]
27. Caddemi, A.; Cardillo, E. A Study on Dynamic Threshold for the Crosstalk Reduction in Frequency-Modulated Radars. In Proceedings of the Computing and Electromagnetics International Workshop (CEM), Barcelona, Spain, 21–24 June 2017.
28. Pozar, D.M. *Microwave Engineering*, 4th ed.; John Wiley & Sons, Inc.: Hoboken, NJ, USA, 2011.
29. Caddemi, A.; Cardillo, E. A low-cost smart microwave radar for short range measurements. In *International Conference on Applications in Electronics Pervading Industry, Environment and Society*; Springer International Publishing: New York, NY, USA, 2018; Volume 512, pp. 41–47.
30. Scalise, L.; Mariani Primiani, V.; Russo, P.; Shahu, D.; Di Mattia, V.; De Leo, A.; Cerri, G. Experimental investigation of electromagnetic obstacle detection for visually impaired users: A comparison with ultrasonic sensing. *IEEE Trans. Instrum. Meas.* **2012**, *61*, 3047–3057. [CrossRef]
31. Richards, M.A.; Scheer, J.A.; Holm, W.A. *Principles of Modern Radar: Basic Principle*; SciTech Publishing: Atlanta, GA, USA, 2010.
32. Di Mattia, V.; Russo, P.; De Leo, A.; Mariani Primiani, V.; Petrini, V.; Cerri, G.; Scalise, L. An electromagnetic device for autonomous mobility of visually impaired people. In Proceedings of the European Microwave Conference, Rome, Italy, 6–9 October 2014.
33. Pieralisi, M.; Petrini, V.; Di Mattia, V.; Manfredi, G.; De Leo, A.; Scalise, L.; Russo, P.; Cerri, G. Design and realization of an electromagnetic guiding system for blind running athletes. *Sensors* **2015**, *15*, 16466–16483. [CrossRef] [PubMed]
34. EM System to Guide Visually Impaired Running Athletes. Available online: https://www.youtube.com/watch?v=GIyQDu0nOww (accessed on 30 September 2019).
35. Hicks, S.L.; Wilson, I.; Van Rheede, J.J. Improved mobility with depth–based residual vision glasses. *Invest. Ophthalmol. Vis. Sci.* **2014**, *55*, 2153.
36. Grant, P.; Spencer, L.; Arnoldussen, A. The Functional Performance of the BrainPort V100 Device in Persons Who Are Profoundly Blind. *J. Vis. Impair. Blind.* **2016**, *110*, 77–88. [CrossRef]
37. Hersh, M.; Johnson, M.A. *Assistive Technology for Visually Impaired and Blind People*; Springer: London, UK, 2008.

38. Dakopoulos, D.; Bourbakis, N.G. Wearable Obstacle Avoidance Electronic Travel Aids for Blind: A Survey. *IEEE Trans. Syst. Man Cybern. Part C Appl. Rev.* **2009**, *40*, 25–35. [CrossRef]
39. Liu, W.; Anguelov, D.; Erhan, D.; Szegedy, C.; Reed, S.; Fu, C.Y.; Berg, A.C. SSD: Single Shot MultiBox Detector. In *ECCV 2016: Computer Vision—ECCV 2016*; Lecture Notes in Computer Science; Springer International Publishing: New York, NY, USA, 2016; Volume 9905, pp. 21–37.
40. He, K.; Gkioxari, G.; Dollar, P.; Girshick, R. Mask R-CNN. In Proceedings of the IEEE International Conference on Computer Vision, Venice, Italy, 22–29 October 2017.
41. Lesecq, S.; Foucault, J.; Birot, F.; De Chaumont, H.; Jackson, C.; Correvon, M.; Heck, P.; Banach, R.; Di Matteo, A.; Di Palma, V.; et al. INSPEX: Design and integration of a portable/wearable smart spatial exploration system. In Proceedings of the 2017 Design, Automation & Test in Europe Conference & Exhibition, Lausanne, Switzerland, 27–31 March 2017.

MDPI

St. Alban-Anlage 66

4052 Basel

Switzerland

Tel. +41 61 683 77 34

Fax +41 61 302 89 18

www.mdpi.com

*Electronics* Editorial Office

E-mail: electronics@mdpi.com

www.mdpi.com/journal/electronics